THE NETWORKED FIRM IN A GL(

The Networked Firm in a Global World

Small firms in new environments

Edited by
EIRIK VATNE
Norwegian School of Economics and Business Administration, Norway

MICHAEL TAYLOR
University of Portsmouth, UK

LONDON AND NEW YORK

First published 2000 by Ashgate Publishing

Reissued 2018 by Routledge
2 Park Square, Milton Park, Abingdon, Oxon OX14 4RN
711 Third Avenue, New York, NY 10017, USA

Routledge is an imprint of the Taylor & Francis Group, an informa business

A Library of Congress record exists under LC control number: 00109571

ISBN 13: 978-1-138-74124-9 (hbk)
ISBN 13: 978-1-138-74121-8 (pbk)
ISBN 13: 978-1-315-18295-7 (ebk)

Contents

List of Figures

List of Tables

List of Contributors

Claes G. Alvstam is Professor of International Economic Geography, Department of Economic and Human Geography, School of Economics and Commercial Law, Göteborg University, Sweden.
Address: claes.alvstam@geography.gu.se

Yuko Aoyama is Assistant Professor and Henry J. Leir Faculty Fellow of Geography, Graduate School of Geography, Clark University, USA.
Address: yaoyama@clarku.edu

Bjørn T. Asheim is Professor of Human Geography, Centre for Technology, Innovation and Culture, Department of Sociology and Human Geography, University of Oslo, Norway.
Address: b.t.asheim@tik.uio.no

Harald Bathelt is Professor of Economic Geography, Department of Economic and Social Geography, Johann Wolfgang Goethe University, Frankfurt/Main, Germany.
Address: bathelt@em.uni-frankfurt.de

Poul Rind Christensen is Professor of Business Economics, Head of the Centre for Small Business Research/Department of Organisation and Management, University of Southern Denmark, Denmark.
Address: prc@sam.sdu.dk

Sergio Conti is Professor of Economic Geography, Dipartimento Interateneo Territorio, University of Turin, Italy.
Address: contise@econ.unito.it

Heikki Eskelinen is an Economist and Senior Researcher, Karelian Institute, University of Joensuu, Finland.
Address: heikki.eskelinen@jouensuu.fi

Paolo Giaccaria is Lecturer in Economic Geography, Dipartimento Interateneo Territorio, University of Turin, Italy.
Address: giaccaria@econ.unito.it

Arne Isaksen is Economic Geographer and Senior Researcher, the STEP-Group, Norway.
Address: arne.isaksen@step.no

Bengt Johannisson is Professor of Entrepreneurship and Director of SIRE), Växjö School of Management and Economics, Växjö University, Sweden.
Address: bengt.johannisson@ehv.vxu.se

Anders Larsson is a PhD student and Research Fellow, Department of Economic and Human Geography, School of Economics and Commercial Law, Göteborg University, Sweden.
Address: anders.larsson@geography.gu.se

Eike W. Schamp is Professor of Economic Geography, Department of Economic and Social Geography, Johann Wolfgang Goethe University, Frankfurt/Main, Germany.
Address: schamp@em.uni-frankfurt.de

Michael Taylor is Professor of Geography, Department of Geography, University of Portsmouth, UK.
Address: mike.taylor@geog.port.ac.uk

Eirik Vatne is Professor of Economic Geography, Department of Economics, Economic Geography Section, Norwegian School of Economics and Business Administration, Norway.
Address: eirik.vatne@nhh.no

Foreword

The aim of this book is to analyse and critically investigate many of the theoretical propositions set out in the debate on flexible specialisation and industrial districts. We have a particular interest in analysing the evolution of industrial systems and the role of SMEs in interrelated networks of firms.

Most of the authors are associated with the International Geographical Union's Commission on the Organisation of Industrial Space. The book draws together economic geographers, economists and management scientists in an effort to produce a deeper understanding of the role of the smaller firm in contemporary economic development.

As editors we have tried to produce something more than a collection of conference papers. To do this we have been dependent on the willingness of contributors to rewrite papers and to write anew for this volume. They have been asked to accept major change to all or parts of their manuscripts. As editors, we hope we have managed to serve their interests and we would like to thank all the authors for their contributions.

Eirik Vatne
Michael Taylor
Bergen
April 2000

1 Small Firms, Networked Firms and Innovation Systems: An Introduction

Eirik Vatne and Michael Taylor

Introduction

The aim of this book is to explore the spatial dimensions of some of the currently more influential concepts in economics, economic sociology and economic geography concerning economic growth. Principal among these are the concepts of *business networks* and *innovation*, the *clustering* of economic activities, the importance of *small firms* in the revitalisation of local economies, and the development of *innovation systems*. Among politicians, for policy makers, and in popular discussion, these concepts are of increasing importance.

Our aim is not to produce a review of fashionable concepts and bold theoretical propositions from which to suggest simplistic policy prescriptions. There are enough of those already, promoting what Lovering (1999) has justifiably criticised as 'policy driven theory'.

Instead, we want to argue that understanding - even of practical problems - is best developed through the construction of theoretical concepts and theoretical explanations. In economic geography as well as in organisation theory and economic growth theory, a deep theoretical understanding has been developed of the processes underlying the creation of economic growth and change in general. Specifically, a large number of theories have been constructed to explain how different economic and social processes create uneven growth among regions and nations. Through these discourses many new concepts have been developed explaining regional economic development. At the same time, the interchange of ideas and models between disciplines has raised new challenges. Likewise, old concepts with roots back to classical economics have been reinvented and

1

revitalised (Martin and Sunley 1996, Martin 1999). Our goal is not to develop more theory, but to bring together different strands of existing theory and to match it with empirical information in an attempt to understand contemporary patterns of economic development.

This is very much in line with the approach to social inquiry suggested by King et al (1994). By drawing together both competing and complementary theoretical approaches to understanding patterns of economic development, we can begin to develop testable hypotheses that can be matched against concrete reality. In this way, we can begin to embark on a process of verification of theory, which is urgently needed in this field.

In parallel with this view, we argue that theoretical understanding is not the same as explaining reality. Theory is a necessary simplification that has to be confronted with empirical data. Only in this way can our understanding on the level of theory be critically examined and brought into line with real life experience. That this confrontation is guided by a transcendental realist or a critical rationalist ontology is not the point at issue. The issue is that theories need to be as concrete as possible. "Vaguely stated theories and hypotheses serve no purpose but to obfuscate" (King et al 1994, p. 20). For too long, empirical tests of theories of local growth have been based on case studies, and usually case studies only of success stories, be they of industrial districts or of business alliances. It is our prime contention that this field of knowledge needs sound, empirical investigation that puts theoretical positions under critical scrutiny.

Already many theoretical propositions on contemporary patterns of economic development, that simplify a complex reality, have become the foundation of frameworks of policy intervention in a world where competition is internationalised and innovation, change and speed is at the forefront of development. There is considerable danger that, in the absence of critical examination, these propositions may create weak, poor and inappropriate policy, much like the growth pole and growth centre policies of the 1960s.

The purpose of this book is, therefore, to match the theoretical debate on patterns of economic development with new empirical insights and critical interrogation. Hopefully, as a reflection of this endeavour, the book will contribute to a refinement of existing theories on the social organisation of modern economies and to a better understanding of the mechanism and the actors involved in industrial dynamics.

The theme

The intention in producing this book is to contribute to the understanding of the dynamics of economic development, particularly in a regional and territorial context. We attempt this through analyses of:

- the functional and territorial *organisation of production*;
- the roles *small and medium sized enterprises (SMEs)* play in production systems; and
- how economic actors (particularly SMEs) relate to their *external environments*,
  - at a micro scale through their *direct relations* with other firms and individuals, and
  - at a meso scale through the influence that context-specific *cultural and social structures* and embeddedness have on their development.

We frame this discussion in the context of *globalisation and change* to explore the extent to which new forms of competition and organisation magnify and intensify the importance of *knowledge, learning,* and *networking* in *localised* processes of development.

Small firms

In economic geography there is strong interest in small firm development for at least three reasons. First, small firms very often are dependent on resources that are embedded in place-based communities. The typical entrepreneurial venture operates in the region where the entrepreneur lives. The social network important for setting up and running the business is also likely to be local. The first marketplace and the first customers are often the other businesses in the region. The local labour market is often a constraint on the kinds of competence the small firm can recruit onto its labour force. In this respect, small firms often reflect the territorial environment within which they have been created, and are more dependent on this local environment than are larger firms.

Second, discussion on the transition from Fordism to post-Fordism has drawn attention to small firms as efficient entities in an economy, well integrated into complex production systems, and with a particular role as flexible and innovative producers. With the production line of the large corporations dismantled, outsourcing and disintegration create new markets for smaller, specialised firms. These transformations are breaking up both the functional and territorial divisions of labour established in previous times, forming new patterns that favour smaller firms, and where geographical proximity seems to matter more than before.

Third, the efficiency and success of localised small firm clusters, has focused attention on the creation of external economies and has illustrated how small firms seem to be able to create economies of scale and scope when they are part of a self-organising local economy. Success stories from 'silicon valleys' and design intensive towns have forced us to rethink the role of SMEs in an economy.

Relational assets

Now, a new heterodox approach to understanding regional economies has emerged, through the writings of Michael Storper (1997a,b) among other, that stands in sharp contrast to the orthodox approach of neo-classical economics. This new approach is centred on the creation and maintenance of relational assets. It breaks down the problem of economic development into three important domains.

First, *technology and technological change* is seen as the principal motor of development, and the capacity to learn and innovate is seen as the dynamic factor of change. To solve complex problems, creative processes are dependent on access to heterogeneous resources from different institutions and dialogue-based social relations. These processes normally take place inside specific local, regional or national territories. Second, economic development is very much dependent on the capacity to mobilise resources needed in the production process and to organise production systems in an efficient and at the same time change-oriented way. This capacity demands the existence of *organisations* - firms or networks of firms - and implies a strong *interrelationship* between firms and various degrees of social and geographical proximity. Third, it envisages the existence of a *territory* where resources are assembled and immobile resources like labour can be mobilised and trained. These are places where interaction between actors can develop without large transaction costs being incurred because of proximity and social coherence. They are places where the collective production of goods can be arranged, and where knowledge externalities are mutually produced and consumed.

Storper (1997b, p.248) has suggested that this is an approach to regional economic development:

> " ... where the guiding metaphor is the economy as relations, the economic process as conversation and co-ordination, the subject of the process not as factors but as reflexive human actors, both individual and

collective, and the nature of economic accumulation as not only material assets, but relational assets."

Network

The metaphor of 'the network economy' is at the heart of this discussion. In economic geography we have arrived at this point through analysis and discussion of new industrial spaces and new forms of the territorial division of labour framed within a new technological paradigm and new regimes of regulation. Regulation theory, coupled with the Fordism - post-Fordism (flexible specialisation) dichotomy, has been a leading theoretical construct that has shaped our understanding of the Third Italy and Silicon Valley and the successful industrial districts of the world. The combination of high economic growth, international competitiveness, small firm formation, local co-operation and the territorial clustering of specific production systems, jointly developed with change-oriented flexible production, has led to the belief that this is a model of industrial organisation for the future. Furthermore, the dismantling of the Fordist enterprise and the decentralisation of decision-making and production in multinational corporations, combined with the growing importance of smaller independent firms, have led us to believe that a new age is emerging based on smaller and more flexible producers.

In economic sociology and organisation theory the network concept has long been paramount. Firm development has been seen as dependent on resources external to the focal organisation. In this literature, the structure of social relations and personal networks is seen as an important avenue for accessing information or exploiting opportunities in the market place. Firms' positions in networks of producers and users appear to enhance their bargaining strength and, more generally, to influence their power as economic agents.

Inspired by Coase and the institutional economists, the firm is also conceptualised as a nexus of treaties. Here the focus is on informal organisation, the flattening of corporate hierarchies and the development of all forms of inter-organisational relationship. Network forms of organisation have also been emphasised in studies of learning, innovation and internationalisation, in terms of the combination of disparate resources in experimental and complementary ways. Lastly, discussions of governance highlight the importance of trust and social relations in developing inter-organisational relations.

In economic growth theory and evolutionary economics, the focus of analysis has turned to knowledge as the most important factor of production, and learning as the most important process of change. This body of theoretical concepts recognises scale economies in the production of knowledge and regards knowledge externalities or technological spillover as economic realities. On the basis of this battery of concepts, this model of economic growth identifies a tendency for economic activities to cluster or agglomerate as a result of direct relationships between firms. In other words, development of any one specific firm is directly dependent on the actions in other firms. Interdependence or networking, particularly in innovative activities, is also postulated as an important part of competition. Furthermore, the significance of externalities includes untraded interdependencies as an important link between firms, very often nurtured and diffused within a specific socio-cultural context.

In total, these theoretical constructs stress relationships other than pure market relationships between firms, and focus on specialisation and the division of labour as important characteristics of modern economic development. In this complex web of relationships, small and medium sized firms are given a prominent position as innovators and flexible units of production. Larger firms are said to be reconstructed into networks of multiple relationships. Subsequently, the two worlds of small firms and large firms are brought together and new forms of collaboration between economic entities are beginning to develop. There also seems to be some agreement that social and cultural forces facilitate collective action and shape socio-economic boundaries between individuals as well as between organisations. On a different scale, territory also appears to play an important role as field of collective action, bringing together individuals, organisations and social institutions. Concepts like 'embeddedness', 'connectivity' and 'reciprocity' reflect this view of interdependence in economic life.

Powell and Smith-Doerr (1994, p. 370) have summed up some of these views as follows:

> "There are essential linkages between economic and organizational practices and the institutional infrastructure of a region or a society. Industrial development need not involve vertical integration or standardized mass production but may rely instead on horizontal network of production. Trust, mutual forbearance, and reputation may supplement and/or replace the price mechanism or administrative fiat."

In this way, the social is brought into the economic, and territory wins a prominent position as one of several explanatory factors promoting economic growth and influencing economic prosperity. It is in this interdisciplinary field of research that the chapters of this book seeks to make a contribution and to explore empirically the hypotheses developed in theory.

Network of relationships

Traditional equilibrium theory has long assumed constant returns to scale and thus that profit is independent of size. Production functions are also specified independently of ownership and, therefore, do not reflect the strategic choices made by firms. Neo-classical economists believe that profits increase with scale up to a certain level, primarily in response to high start-up costs. Thereafter, profits decrease with scale owing to complexity and the difficulties of management and control in very large firms. According to thinking in industrial economics, there should be a minimum efficient size of firm owing to the indivisibility and 'lumpiness' of capital equipment. This size might differ between industries, but a firm would still have to reach a 'critical mass' to be competitive.

In the strategic management literature, a common belief is that the relationship between market share and profit is U-shaped. Large firms can win cost advantages through scale economies and learning effects. Small firms can differentiate production and target small segments in the market. At the same time, this literature warns against a middle of the road solution through diversification. In theories on the growth of the firm, a static view such as the one just presented would be rejected. It is the growth of the firm, not its size that results in higher co-ordination costs and lower profits. Stochastic growth theory, however, suggests that it is a stochastic process that distributes firms in different size groups. Meanwhile, management theory would support the view that there are often constant returns to scale in many industries, but that management may want to expand company size even though that may not necessarily generate maximum profits for the shareholders.

In this book, we argue that the disadvantages of small scale can be overcome by networking and that, at the same time, some of the advantages of small scale can be expanded through the building of inter-firm relationships. The term 'network', however, has many meanings and is very often used as an analytically empty concept. It is frequently used as no more than a synonym for informal ties inside organisations or for any linkages and relationships between organisations (see, for example, Dicken and Thrift 1992).

The 'network' as an analytical concept

As a more strictly defined analytical tool, the network concept is used in many disciplines of research. In the social sciences, network analysis is used:

- to analyse a specific form of governance;
- to portray the external environment of an organisation; or
- as a formal research tool for analysing the power and autonomy of organisations.

For some, a network form of governance is a *hybrid form* for organising transactions. It is not governed by pure market mechanisms and price, and neither is its transactions governed by a hierarchy of larger companies. It is indeed a hybrid form of bilateral or trilateral governance brought about by the need for transaction specific investments, customised purchases or low transaction frequency (Williamson 1979). It can be regarded as a metaphor for inter-organisational relationships and different forms of co-operation, from joint ventures, controlled franchising or technology financing, to different forms of system integration as seen in industrial districts or other production systems (Thorelli 1986). In these circumstances, the focus is often on the individual firm or dyadic relationships. What is important, however, is that this type of relationship brings the social world into the economic.

Others see networks as *highly specific forms of governance* based on social relations, trust and the sharing of complementary resources. They are a combination of market and bureaucratic forms of governance where both authority and market solutions influence transactions (Powell 1990, Burt 1992). The focus is not on individual firms or dyadic buyer-supplier ties. Instead, it is on the network as a whole. It is the sum of the interrelationships and the complementary resources that are brought into the network that fuel the system with competitive power. Industrial networks could, in this way, be analysed as *aggregations of relationships*. By this meaning of the term 'network', relationships are a form of long-term and general inter-firm behaviour that has five elements (Johanson and Mattsson 1986):

- *mutual orientation*, or a willingness to co-operate and interact with each other. An important reason for this willingness is the complementarity of firms' objectives. It facilitates the more effective acquisition of resources, and the distribution and sale of products. It also promotes specialisation and the division of labour. As a result, costs are reduced while scale economies, the joint use of resources, and innovation and sales are increased;
- *dependence in relationships*, which intensifies over time bringing problems of power and control. In turn, asymmetries in power are related to different parties' control over resources that are valuable to others;

- *bonds* between firms will also develop over time into ties that can not easily be broken. With time, the different parties in a relationship can invest so much that they have a vested interest in keeping it going. In this way strong bonds provide the network with a more stable and predictable structure, but not necessarily a static structure;
- *relation-specific investments* are also created involving both hardware and software. They can be idiosyncratic in nature and not easy to use or regenerate if the specific relationships break down;
- inherent in every relationship is the tension between conflict and co-operation. This tension creates the *atmosphere* of the relationship and affects the trust between partners. It influences the openness of the dialogue of the relationship as well as the sharing of resources.

Easton (1992) suggests that networks can also be analysed as structures, positions and processes. Networks form *structures* because firms in industrial systems are interdependent with other actors inside the system. Behaviour is, therefore, in one way or another constrained, and the firm is *not an atomistic unit* operating in a pure market economy. Obvious reasons for the formation of industrial systems are the quest to develop specialisation economies and the need to match heterogeneous resources within a horizontal or vertical division of labour. These reasons are reinforced by changes in demand, especially the increasing demand for heterogeneous products. Through specialisation, different relational adaptation processes, learning processes and transaction-specific investments, interdependence develops among actors in a network. This is interdependence based not on homogeneity and the utilisation of traditional scale economies, but interdependence based on complementarily and the division of labour. It is interdependence based on *heterogeneity*. Heterogeneity also means that network structures are flexible. They can adapt to different demands and they can combine resources in many different ways. This implies that we should not look for a static optimal network structure but accept that networks can be organised in different, equally efficient combinations, and will be reconstructed and reformed as part of the ongoing processes within the system.

The structure of such a system could take different forms ranging from monolithic networks of a few large firms to a myriad of small firms taking part in the production system of an industrial district. It could consist of a dominant large firm and its tightly structured sub-contracting system, or it could take the form of a self-organising system and a loosely coupled network. Given their variety and fluidity, it is clearly very difficult to sharply define the boundaries of a network.

A network constrains and circumscribes the action field of participating firms, but its also opens up new opportunities for accessing external resources available inside the network. A firm's *position* inside such a network will define how the balance between constraints and opportunities is revealed for the focal firm. Power could be said to lie in the structural position of a specific agent. Inside structures, firms will act as nodes, occupying a particular location in the structure with direct and indirect links to other nodes. Any firm could assume the role and position as a 'hub', a 'carrier', or an 'isolate', for example. Its position in the network could be measured in terms of autonomy and constraints. Generally speaking, empirical studies seem to suggest that centrality in inter-organisational networks correlates with power.

To summarise this discussion, inter-firm relationships in business networks display the following characteristics:

- at least three parties must have some sort of commercial relationship;
- each of these firms is dependent on assets controlled by other partners in the network;
- partners have some degree of independence at the same time;
- network relationships need transaction-specific investments from both parties - investments which are semi-specific in character;
- it takes time to develop network relationships;
- a firm can be part of more than one network;
- different power structures can be identified in networks. One generic model identifies an asymmetric power structure where a hub firm dominates the network. Another model is based on a more symmetric balance of power between partners;
- inside a network, there must be incentives available to govern exchanges. Agreements rely on negotiations and consensus;
- the management of networks will be organised according to the strategic interests of the partners and the structures of power involved. Again we can identify two generic models: a formal economic arrangement based on self-interest, and a more organic form based on trust and behavioural adaptation.

A localised process

Networks help to match supply and demand and to transmit, through personal communication, a lot of information not accessible on the open market. As argued by Granovetter (1985), strong ties in a network are an important aspect of trust building, that serve to develop mutual dependence and collaboration. Strong ties can, however, limit and close the circulation of information, leading to the reproduction and distribution of the same information in a closed circuit. Weak ties, in contrast, are important for access within networks to new information or other resources. Networks of only strong ties are closed circuits that soon stagnate, reducing their internal capacities to innovate and develop competitive capabilities.

Burt (1992) claims that it is important for a focal firm in a network to construct a structure of non-overlapping relationships. If the purpose of a network as a whole is to monitor, access and share information on markets or technology, firms are in a position to economise on those relationships. Firms can concentrate on their primary contacts alone in a cluster of interrelated firms. The rationale here is that one central contact is sufficient to secure most of the information circulated in that cluster. Firms with limited resources should, therefore, link up with and tap into other clusters, instead of linking with other members of the same cluster.

These propositions suggest that closed local industrial networks are not necessarily the most efficient or innovative mode of industrial organisation as suggested by the most eager promoters of the industrial district model. In the literature of regional science and economic geography, it is territorial context that is emphasised, with proximity and locality being vital for economic development. Network theory, in its less deterministic form, recognises that other kinds of organisational arrangement can be as effective, innovative or adaptive as the industrial district model.

Generally it is argued that inside networks competition in the neo-classical sense is replaced by rivalry for the control of resources, and that co-operation and rivalry are combined dialectically. Inside networks, the daily co-ordination of processes is not guided by a master plan or quasi hierarchy, but appears to be arranged in a self-organising way. Network processes are controlled by the internal distribution of power in the networks and the network's structural layout. As a result, there will always be tensions and latent conflicts of interest between network partners, creating turbulence and change. Industrial networks are, therefore, stable but not static. Participants come and go depending of the resources they need and the power relations of the actors involved.

It can also be argued that continuing relationships between firms create the necessary environment for the development of trust and learning, and the give and take of collaboration. Time is needed to develop social exchanges, and time provides the opportunities for experimentation and open communication. A lot of 'timeless' studies of the territorial division of labour and localised innovation processes remind us that there is an urgent need to introduce time in these analysis (Massey 1997).

The complementary nature of innovative collaboration necessitates the sharing of resources built on the strategically important core skills of the participating firms. In the creative phase of innovative activities, face-to-face and dialogue-based communication is needed. An important part of these exchanges involves tacit knowledge. To generate new products in a network requires available knowledge resources to be mobilised and organised in the face of uncertainty.

There are strong arguments that this form of resource mobilisation occurs best in localised settings:

- in regions able to mobilise heterogeneous resources from both private and public institutions;
- in an environment where social institutions and cultural structures promote entrepreneurship and help to lower transaction costs;
- where investments in knowledge takes place even if part of the investment leaks out as a collective good to other, local firms;
- where the investing firms benefit from technological spillover effects produced by other participants in the network; and
- where there is access to knowledge capital.

In short, regions can function as the setting in which physical, human and social capital can be combined to create a highly competitive environment where high productivity, innovativeness and adaptability co-exist. The studies presented in this book reflect on and 'test' these propositions. At the same time, we shall ask if the local and regional scales will always be of importance. Under what circumstances will the national or international scales be the most important environments for firms' external relations? Network theory also tells us that social relations are the most important channels through which information flows. Social relationships are, in practice, the glue in the network. We can ask if it is possible for trustworthiness and open communication between partners to develop over time in the absence of geographical proximity? If so, under what circumstances will space matter? An important task of this book is to challenge the 'territorial' version of the model of economic development outlined in the earlier sections of this chapter, in a search for a deeper understanding of the influence that territory has on economic development.

The book contributes to the analysis of *relationships inside networks* and to a more sophisticated understanding of the division of labour and the function of different types of firms in networks. Further, the role of *geographical scale* and the relative importance of local versus global processes are questioned. In different chapters we also challenge the understanding of networks as timeless, static structures. Important *processes* involving power relations, learning dynamics and lock-in mechanisms develop over time and influence the formation of networks and their territorial form. Lastly, the roles of government and policy in innovation and networking are examined.

The structure of the book

The book is divided into three parts. In Part One, the first two chapters serve to introduce the subject of the book - small firms and their role in industrial systems and economic development. Part Two comprises four chapters and analyses the evolution of production systems organised by large firms. The separate contributions emphasise the challenges facing smaller sub-contractors when supply chains are restructured. In Part Three, five chapters investigate the importance of various aspects of local resources and local networking for the functioning and performance of smaller firms. In particular, they explore processes and forces that might modify, undermine or even destroy the self-organising systems of industrial districts. The separate chapters explore issues surrounding knowledge, learning, local resources, exogenous influences, power and managerialism, 'testing' the industrial district model in concrete empirical contexts.

In Chapter 2, Conti outlines the emergence of the institutionalist debate on the role of small firms in economic development. He begins with a critique of industrial dualism and its neglect of small firms as important actors in dynamic industrial systems. The key argument is the need for variety and difference in economic processes and the potential that smaller firms offer in this respect. SMEs are a crucial element in all the new models of economic growth that have been developed against the backdrop of recent economic change, restructuring and instability. Conti elaborates the pedigree of 'network' ideas and the functioning of networks: as localised production system and self-organising systems; as a place for collective learning; and as producer of externalities and social capital. In particular, he emphasises the knowledge and learning that arises from the social construction of local economies, together with the mechanism of trust, reciprocity and loyalty that underlie them. He also identifies a need to create an appropriate 'network' policy to foster local economic development.

Aoyama takes up this policy thread in Chapter 3 to analyse the relationship between policy and the formation of industrial networks in Japan. She spells out in detail the fallacy of taking the Japanese lean production system as a model of 'learning' and reciprocal, networked relationships. She shows the Japanese system of small firms to be an artefact of policy – forced, government controlled collaboration for the benefit of large corporations. In the current long-running economic crisis in Japan, the sub-contractor system built on small firms does not appear to have produced the benefits it was said to able to create from reciprocal, trust-based relationships between firms.

Christensen adds to this critique in Chapter 4, drawing on Danish experience. Traditionally, sub-contracting has been viewed as a regional or national phenomenon. Development towards global supply systems challenges the way sub-contracting systems are organised. This study recognises that sub-contractors have internationalised and grown in size to meet the demands of globalised principals. Multi-sourcing is replaced by single-sourcing. A segmentation of supplier relationships is identified and a new typology on sub-contracting is developed. Christensen recognises the potential for local growth from trust-based systems, but sees them under threat from processes of globalisation. Those trust-based relationships also start to become non-local and to be stretched overseas as sub-contractors internationalise to match their principal contractors, and TNCs tap into localised clusters and impose their managerial culture on smaller firms.

In Chapter 5, Giaccaria adds to the analysis of the transformation of localised sub-contracting systems through an analysis of the transformation of the Fiat/Turin production system. He identifies two distinctive development trajectories. First, a 'post-Fordist mesosystem' has developed out of the traditional, Fordist sub-contracting system centred on Fiat. This system organises external relationships at a global scale while simultaneously taking advantage of the territorial division of labour. Internationalised companies now co-operate at the global scale and, as a consequence, break up the traditional local sub-contracting system. Second, a 'local value production system' has emerged based on development and engineering skills that have accumulated in the Turin area. It is still a local network system, but it now serves international markets and companies other than Fiat.

Extending this analysis in Chapter 6, Alvstam and Larsson demonstrate through a detailed analysis of Volvo-Göteborg's sub-contracting system, the same trends of concentration and internationalisation among the firm's suppliers. In a study of firms in the supplier park adjacent to the assembly plant, this chapter shows that large corporations are not really capable of building 'look-alike clusters' of sub-contractors that can re-create the conditions of networked trust, reciprocity and loyalty from which the dynamic, innovative capacities of SME-based industrial districts have been said to derive. Corporate sub-contractors develop no local collaboration that might inspire 'learning' and instead remain tied to their parents in other locations. This analysis significantly undermines the idea of using corporate branches to kick start learning processes in backward or declining regions.

Finally, in Chapter 7 in this section, Bathelt provides a further perspective on the de-concentration of the Fordist production systems of large firms and the reintegration of SMEs into those systems. He does this by analysing the production processes and structural changes in four branches of the German chemical industry. Using cluster analysis and qualitative data, he identifies three distinct sets of chemical firms: a group of semi-flexible, partially integrated firms; a group of flexible, specialised firms; and a group of conventional, specialised firms. The analysis shows that the restructuring of formerly Fordist production systems has not followed a uniform pattern. However, most of the surveyed firms have not opted to increase product and process flexibility. And, the social division of labour has also remained relatively fixed. The study does, however, show that technological spillover effects have occurred and that geographical proximity is important for some inter-firm relationships, at least in the chemical industry. In other words, while change has been accompanied by the retention of some Fordist rigidities, it has also involved in certain situations more flexible interconnectedness.

In the first chapter of Part 3, Asheim and Isaksen (Chapter 8) detail the most recent thinking on regionalisation and regional innovation systems, and try to develop an explicit link between theory and policy. They begin the chapter with a critique of the 'learning' model of industrial districts. They see it as having been applied too widely and having been derived principally from success stories. They also suggest that the emphasis on tacit as opposed to codified knowledge has been over drawn. They argue that codified knowledge may well be necessary for tacit knowledge to be mobilised and used effectively for anything more that incremental innovation. The case studies they present demonstrate the dangers of over-emphasising the learning model in some circumstances. But, throughout the empirical discussion, the role of government in interventions between firms and institutions is constantly to the fore, although it is rarely dealt with explicitly and built into industrial district theory.

In Chapter 9, Taylor suggests that the unequal power relationships between firms in networks will limit, and limit differently according to the power firms have, their ability to participate in local learning systems. Drawing on empirical evidence form Australian firms, the transaction patterns of different types of firm (more powerful/less powerful) are seen as being constrained by their 'statuses' within power networks. As such, some firms gain more from local systems than others do. It is suggested that, over time, any symmetrical relationships that might exist in a cluster will be eroded. Clusters, therefore, contain the seeds of their own destruction.

In their analysis in Chapter 10, Eskelinen and Vatne challenge the proposition that the mobilisation of external, local resources is an important determinant of the success of small firms in international markets. In this study, Nordic SMEs are grouped into three types: non-exporters; experimental exporters; and committed exporters. They are analysed in terms of their access to internal and external resources, their co-operative relationships, and the territorial environments or milieus within which they work. The findings suggest that most SMEs produce and export simple products with little need of external resources. Internationalisation is best 'explained' by the firm's access to internal resources. Local production systems and local resource endowments do not 'explain' exporting. On the contrary, the most successful exporters co-operate more with their foreign partners than with local firms.

In Chapter 11, Schamp also suggests that for clusters to remain efficient, influences external to a network may be just as vital as internally generated trust, co-operation and collaboration. These external connections breathe new ideas and new practices into old areas. Nevertheless, although they may help to secure continued economic activity in a cluster, they may also be symptomatic of that cluster's long-run decline. External influences can, therefore, both bolster and destroy local systems and clusters. These conclusions are based on a longitudinal study of the evolution of two traditional industrial districts in Germany based on footwear and wickerwork production.

Finally, in Chapter 12, Johannisson extends the analysis of the potential destruction of the self-organising systems of industrial districts brought about by invading larger firms and the introduction of well meant industrial policies. He argues on the basis of Swedish evidence that the processes of managerialism (including planning interventions, policies and programmes) undermine trust-based relationships and, therefore, destroy the learning capabilities of clusters and local areas. In other words, the exogenous interventions of large firms in clusters, the evolutionary changes that foster larger firm in agglomerations (i.e. power), and government promotion of regional innovation systems, may actually destroy region-specific 'learning' systems.

PART I

SMALL FIRMS VERSUS LARGE FIRMS: EFFICIENCY, FLEXIBILITY AND INDUSTRIAL POLICY

2 Small and Medium Sized Enterprises in Space: The Plural Economy

Sergio Conti

Introduction

The issues of corporate geography, which have dominated debate in economic geography for decades, have arisen in a precise historical setting. The debate has focused on the inevitable rise of large-scale corporations and from their growing influence in the world economy. At the same time, it has reflected an ideology of development built on the choices and behaviour of large companies, which were reckoned to be capable of integrating and controlling space and exporting their own rationality. In this framework, the *region* was defined in relation to exogenous forces (major companies), which were seen as the only forces capable of structuring production, overseeing the diffusion of technology and activating processes of social and spatial interaction (centre-periphery, dominance-dependency).

However, economic upheaval at the end of the millennium has tended to challenge this interpretation. First, regions and places are now seen as complex socio-economic entities whose organisation and dynamism are not separable from inherited history and cultural heritage. Second, the inability of orthodox models (polarisation, spatial division of labour, cyclical and sequential models) to explain the 'new' phenomenon of industrialisation in previously non-industrialised regions that are not dependent on the traditional engines of development has been recognised. Third, a complex dialectic between company behaviour and space has been acknowledge which opens up the concept of an *intermediate entity* - an entity between the system as a whole and the individual economic actor. The identification of the *local system* as the point of reference of theoretical reflection and empirical research shows in particular the fertile consequences that stem from the adoption of this critical point of view.

As a whole, these changes have thrown into disarray old logic on the organisation of knowledge. It can be argued that now a different logic and ethic of knowledge has been constructed that has had a profound impact on the social sciences. The change is marked by the rediscovery of the *small enterprise* as a key actor in new and intense processes of economic and industrial development.

The origins of industrial dualism

In the tradition of the social sciences, the search for contrasts and differences in the economy is largely found in the debate on *industrial dualism*. In the economics literature after the Second World War, the concept of dualism was used to explain the simultaneous presence in a country of two different economic sectors. The first sector was advanced and underpinned by capitalist relations of production. The second was traditional and backward and characterised by economic structures and relations of a pre-capitalist type. This vision reflected divisions among companies as well as in labour markets and inspired at the same time a more general approach to understanding the relationship between developed and under-developed economies.

It is worth recalling that the thesis of industrial dualism derived from the rather elementary observation that each industrialised economy can be divided into large and small production units. This distinction was first introduced by Miyakawa (1964) in his studies on Japanese industrial structure and was definitively re-ordered in the late 1960s when Galbraith (1967) and Averitt (1968) independently reached similar conclusions on the industrial dualism present in the US economy.

It would be inappropriate here to examine these pioneering analyses in too much depth, just as it would be inappropriate to examine the trade categories (quite often inadequate) which were the basis of them. From our point of view, it is enough to recall how these contributions opened a lively debate amongst economists. The challenge to orthodoxy that they prompted developed on two fronts: the first emerged from the cradle of neo-classical thought (Penrose 1959); the second had cultural roots that were to varying degrees Marxist (Santos 1979, Attali 1975).

Building on enterprise dualism has allowed us partially to overcome the rough outlines of the early studies on economic dualism (with the partial exception of those of Edith Penrose). In the neo-classical conception of equilibrium, the contrast between the two sectors (capitalist and pre-capitalist) was held to be an accidental and temporarily disturbing phenomenon that would be resolved spontaneously with the spread of the features of 'central' development to the periphery. The duality of the system is now represented, in contrast, as the natural character of an economy which always contains within it different sectors and actors, each of which plays an essential role in the functioning of the economic system.

On a different front, the theses of dualism implicitly overcame the orthodox Marxist vision of the irreversible historical evolution of the forces of production, the cause of the gradual decline of 'residual' organisational forms (the small enterprise in its various forms, in fact). This essential and deterministic view was now countered by the conception of an industrial universe structured by relationships of dominance and dependency between two distinct sectors present in the system; the small enterprise as a functional reality within an hierarchical structure dominated by large-scale capital. In reality the unsolved problems of orthodox thinking had been evaded but not resolved. The implicit superiority of modernising forces in the economy were acknowledged, represented by large corporations, while the co-existence of small enterprises was admitted which allowed a less ambiguous definition of the features and dynamics of the dominant segment. Thus, the system was divided into two contrasting segments, neglecting the complex reasons that regulate their dynamics and the great variety of the relationships that are established between actors in an economic system.

The small enterprise in the contemporary economy

Economic systems do, in fact, have many more discontinuities. As Berger and Piore (1982) maintain, dualism does not necessarily imply a division into *two* autonomous and discontinuous segments, but the fact that *a society is divided into segments and is not organised in a continuous fashion.* The fact that there are two or more blocks of this kind is not fundamental. The important thing is that the number of segments is not multiplied indefinitely, otherwise a continuum would form again.

In their turn, in a series of essays published in the early 1980s, Taylor and Thrift (1982a and 1983) propose an alternative to the framework outlined above, with the evident purpose of providing a more suitable scheme of reference. In their seminal work, the process of segmentation of the enterprise system is assigned an historical nature that can thus be traced back to broad political, social and economic conditions. Each new configuration would contain new segments, added to the existing ones, which would explain the relations of domination and subordination present in later configurations. It follows that each 'model' of articulation stems from its preceding structure and is at the same time the origin of its later structure, which become increasingly more complex and interconnected (Taylor and Thrift 1982a, p. 451).

The works of both Berger and Piore and of Taylor and Thrift had a great impact on geography in that period, suggesting the need to get to the roots of how a real economic system works. Their reasoning has to be put into the economic context of the time (the 1980s), when the phenomenon of the small enterprise began to acquire a completely new meaning. In the previous three post-war decades of constant economic growth in the developed countries, most economists, politicians and company managers had not hesitated in putting forward the idea that small companies represented an archaic structure, partially condemned by the very evolution of industrial capitalism. This attitude was quite unsurprising because in those immediately post-war decades, small companies' share of production had gradually shrunk, and numbers of small companies had fallen. In the 1970s and 1980s, however, arguments used to support company expansion (such as the effects of scale, the effects of learning, and the effects of variety) changed from being factors of competitiveness to elements of vulnerability. This brought into question the industrial rationale of the post-war period together with the 'great certainties' on which the success of the international management schools had been built.

Economic reality changed profoundly in that decade, and the rise of the small enterprise in many countries was described as 'a full-scale break in the natural evolution of capitalism' (Julien and Maurel 1986). Its rise was helped in part by the general crisis of the Fordist-Taylorist model of industrial development which, laying bare the limits of large size, highlighted instead the advantages of the small company with its greater production and management flexibility.

A new reality emerged which cannot be explained by contingent events or limited to individual countries or production sectors. That reality has three main dimensions. First, there is a growing segmentation and *variability of market demand*, which brings into question corporate efficiency and places the small enterprise, with its higher production efficiency, in a position of relative advantage. Second, there are shifts in *technological development*. Production that had previously demanded high capital investment levels is now also possible in smaller units, radically changing the link between size and productivity. The increasingly complex nature of innovative processes implies, in turn, the involvement of many actors (companies belonging to various sectors, and public and private institutions), again assigning a central function to small enterprises. Third, there is a new labour market dynamic. Rising wage levels and growing labour market rigidities have encouraged the decentralisation and relocation of production. *Labour market segmentation* has limited people's employment opportunities, encouraging young people and women into self-employment, generating widespread forms of diffuse, small-scale enterprise, at least in highly industrialised countries.

Numerous quantitative analyses have demonstrated how, since the 1970s, SMEs have played a non-trivial role in the creation of new jobs, determining in many cases the fates of regional economies. In almost all industrialised countries, it has been this type of company that has, proportionally, generated higher shares of employment than large companies. These small firms have distinctive characteristics though they are hard to define. First of all, it is necessary to remember how hard it has been to reach a comprehensive definition of the concept of the small company. Small companies embrace a heterogeneous range of actors, employing different technologies, working in different sectors, using different operational logics, and having high rates of entry and exit. They vary substantially in size, whether size is measured as added value, assets, turnover or numbers of employees. In the United States the Small Business Administration usually classifies as small companies those with fewer than 250 employees. In most European countries (the exceptions are Austria, Norway and Switzerland) and Japan this category covers companies which employ fewer than 50 workers (Julien and Marchesnay 1988, Storey and Johnson 1986, Bolton 1971, Ciampi 1994).

There is at the same time substantial agreement that in qualitative terms SMEs share common relations with their operational environments. Their organisational configurations tend to be distinctive, with centralised and personalised decision-making structures, and their management structures are often simplified and intuitive (Laufer 1975). Relationships with their task environment are particularly important, including links with suppliers, customers, competitors and, more generally, the information retrieval system (Thompson 1967). It is now realised that the small enterprise is not an exception (or an imperfection, according to the neo-classical orthodoxy) in the economic and social world, but a fundamental aspect of the way in which a society organises itself and produces.

The theoretical lesson is perhaps even more important: the debate on small companies has stimulated a profound review of the social science discourse and economic thinking on capitalism's real essence: the role of variety and difference (between countries, regions and companies) as key dimensions of economic processes. This is the key to the analysis presented in this chapter. Alongside the rise and consolidation of complex global company networks, the emergence on to the scene of the small and medium size enterprise has been crucial in the search for new meanings in a period of radical economic change. This chapter reviews the main stages in this debate on economic change. The debate is a series of passages that are scanned very briefly, as they are well known in the territorial sciences. The presentation of the debate is necessarily partial and fragmented owing to the complexity of the discourse. For this reason, the chapter concentrates on summarising the logic of the debate and, above all, the main conceptual innovations that have emerged.

The contexts: the decentralisation of production and the 'peripheral' economy

Initially, the transformations in industrial structures inspired theories on the *decentralisation of production*. This was the beginning of a stream of research which had considerable impact at theoretical and operational levels, and whose most significant result was the breaking down of the system of companies into a plurality of divisions not brought out by traditional conceptual schemes. It is still true that in this period the relationships of domination and subordination between large and small enterprises was again underlined as a phenomenon that was *organic* to the functioning of the economy. However, the explanation no longer stemmed, as before, from the blinkered vision of a world that could be divided into two clearly separated segments. Small companies appeared, on the contrary, as a structural element in the mechanism of accumulation thanks to their articulation in segments, each of which was assigned in its own function. A rigid, causal subordination was thus replaced by a functional and typological articulation, with the small company ceasing to be considered as a 'morphologically' homogeneous reality.

The studies of Italy's peripheral economy (the most evident case of industrial dualism in the West, according to Berger and Piore) have been much more significant, however. These studies highlighted the incompleteness of analyses of the decentralisation of production and the image that they gave of small companies as dependent or at the most involved in residual or niche production. It was revealed that significant numbers of small companies did not conform to these traditional operational parameters. First, sets of SMEs possessed their own autonomous markets (national and international). Second, it emerged that many of these companies followed a locational logic significantly different to that of the past. Instead of the regions that had written the history of Italian industrial development (the north-west), they preferred as locations other areas (the small and medium cities of central and north-eastern Italy) that specialised in particular types of production and offered significant agglomeration economies.

Since then, criticism of the dualistic thesis, that had been used to explain north-south imbalances in Italian development, has been accompanied by criticism of other dualistic visions that have also obscured a more complex reality. Those dualisms include development and underdevelopment, advanced regions and backward regions, and small companies and large companies, for example. In the interpretation of the economy, therefore, there has been a shift from a *unique* model, coinciding with the large company and its mechanisms of development and regulation and based on the contrast between development and underdevelopment, to the simultaneous presence of *multiple* models of the development and transformation of regional economic structures. From a sectoral perspective, these images overlap: in the specialised sectors of the 'core', 'central' parts of the Italian system (characterised by large companies in transport, chemicals, household appliances, electrical and electronic consumer products etc) Italy's share of exports from industrialised countries has fallen. In contrast, competitiveness in Italy has grown in the sectors, both traditional and modern, typical of the 'peripheral' economy.

The Italian situation exemplifies a phenomenon common across all Western economies. The new economic realities that have become evident have brought into question the criteria used in traditional theoretical analyses and have stimulated attempts to explain the new forms of organisation of the economic system and the development of new explanatory concepts. The simplest way to explain the industrial, political and economic transformation that originated in the 1970s was initially to turn to technology. Indeed, it is undeniable that the introduction of new technologies, especially process technologies, rapidly made the organisational forms of the past production systems obsolete. But, this was a simplistic approach that, in effect, reproduced the old rules of scientific analysis. The introduction and generalisation of new technologies is not, in fact, a phenomenon that occurs by chance at a lucky moment in history. If the years we are discussing were a period of deep crisis for Western economies, it would be unreasonable to uphold *tout court* that the new technologies were the origin or cause of the crisis. Observers began thus to wonder about whether new technological and organisational solutions should not instead be interpreted as a response to the contradictions that the crisis had revealed in the capitalist system that had been based for years on taken-for-granted certainties. The idea thus began to emerge of a close link between the two phenomena - the arrival of new technologies and the transformation of modes of organisation and production - which only history was able to clarify. And a circular relationship reappeared, linking the categories of reality and their historical development that had been neglected for a long time. From many points of view, this marked the revenge of the historical dimension in the explanation of the economy. It was recognised that criteria for the interpretation of reality required a substantial change in the standpoint from which that reality was observed. If, in fact, we rapidly review the succession of ideas that have asserted themselves since that period - as we shall do in the following pages - we discover that to better understand the breaks that appear in reality and in related conceptual categories, a key priority for interpretation stems from the assumption of *historical* discontinuity in the development of our society.

The lesson of history: mass production and flexible production

Few books have had a greater impact in the recent debate within the economic and geographical sciences than the work of Piore and Sabel (1984). This book raised the small enterprise to the status of a significant theoretical category. The book is also important because it interpreted the rise of the small enterprise as a major divide in the development of Western economies. This explanation derives from a particular interpretation of history, seen as a periodisation of successive and differing phases. In the 1980s, the success of the small enterprise was not viewed as passing or contingent, but as the emergence of a new economic paradigm that replaced the Fordist-Taylorist logic that had underpinned the formation and consolidation of large companies in the past. Piore and Sabel's reasoning was inspired by regulationist theory and the principles of economic institutionalism. The new system of flexible specialisation (of the production process and the labour market) was, in effect, seen as an alternative to mass production: resting on the competitiveness of non-standardised goods and on the organisation of production based on production units. In its turn, technological development, far from being a routine process like standardised production, was seen as a continuous activity, managed through co-operation between actors. All of this was used to explain the rise of institutional structures, which formed production systems that greatly resembled those that dominated the industrial world before the advent of mass production (Sabel 1989).

Piore and Sabel's thesis is not, however, free from criticism (see, for example, Martinelli and Schoenberger 1992). Proposing flexible specialisation, as a general model alternative to the old Fordist model, ends up masking the greater complexity of the system, where several models can co-exist without reciprocally ruling each other out. It is not, in fact, by turning to a deterministic - and indeed romantic - interpretation that one can answer the questions posed by practical and theoretical problems. Nonetheless, this was a turning point in economic culture that, by getting behind the apparent chaos of reality, laid bare the limitations of the hypothesis of the intrinsic superiority of the structures and processes that had previously dominated the century's economic scene.

Naïve illusions: transaction costs and geographical industrialisation

There has been, therefore, a 'weakening' of the role of the corporation in the economy, the rise of new models of growth based on SMEs, and an explosion onto the scene of 'new' and dynamic production spaces. Against this background, a major theoretical issue has been to clarify the ambiguities in the concept of a transition from the 'organised capitalism' of Fordism to alternative flexible organisational forms. In this direction, one theoretical attempt came from the rediscovery of *transaction costs* as a mechanism to achieve the efficient management of production (Scott 1983 and 1988b, Storper and Walker 1989, see also Moulaert and Swyngedow 1989). According to this theory, the structure and boundaries of a firm can be understood as a trade-off between the cost of governing the relations (transactions) that can take place either in the market (social division of labour) or within the company itself (technical division of labour). In other words, there is the choice between co-ordinating production through internal transactions, with the consequent increase of the organisation's operating costs and the size of the company itself, and allowing these transactions to be guided more or less freely by the market.

Given these premises, the territorial organisation of a production system is interpreted as stemming from the dynamic of the functional interactions between the company and the context within which it operates. This is a significantly different interpretation of the territorial organisation of production to those of classical and behavioural formulations of industrial location. If company organisation is a consequence of the striving for savings in production (which are achieved by strategies of vertical integration and disintegration), the same striving can also explain why large size may not always be the most efficient organisational form, and why a significant sector of small companies develops. In the new industrial landscape, the technical indivisible and vertically integrated structures of the past are thus replaced by a 'mosaic of agglomerations', or vertically disintegrated production complexes, which represent the organisational form around which the dynamic of a flexible production system pivots.

This theoretical proposal, consistent with the regulationist hypothesis and the thesis of flexible specialisation, is antinomic to that which inspired the formation and consolidation of vertically integrated company structures. In the case of vertical disintegration, in fact, the scenario is overturned. Flexible specialisation is accompanied by the elimination of obstacles to the establishment of new organisational models: it is as if, almost by magic, cultural forms and types of social interaction, organisational practices, interdependencies regulated by competition and co-operation could again deploy themselves freely. In practice, in highly technological districts and in areas in which an ancient craft tradition is revitalised, the Fordist ethic is no longer at home. What is more, by focusing attention on costs, the transaction costs approach makes it difficult for the greater complexity of economic phenomena to be appreciated. The abstraction involved leads to the issue of *identity* (of actors and their spatial contexts) being ignored, even though this is a fundamental aspect of historical continuity.

In the geographical application of the transaction costs theory there is a major theoretical problem: a fundamental methodological reductionism that can be traced back to the fact that the market is assumed as the fundamental organising principle. More precisely, the relations between economic actors are simplified and ascribed to a generalised search for efficiency, which is pursued through the minimisation of transaction costs. As Grabher (1993a) notes, the institutional context underlying business decision-makers' actions and the greater variety of relations that develop in a competitive market are ignored. By removing the company from its social context, the transaction costs theory fails to come to grips with non-trade relationships between firms, which are increasingly important in the functioning and interpretation of the contemporary economy.

Marshallian external economies and the rediscovery of civicness: between modernity and tradition

A completely different way of understanding the processes forming 'new' local production systems has its roots in the rediscovery and re-evaluation of the work of the British economist Alfred Marshall. A sizeable group of contemporary economists and geographers (Becattini 1978, Brusco 1989, Sforzi 1989, Conti and Julien 1991) have turned to the work of this classical economist because of the need to find a theoretical point of reference capable of explaining phenomena empirically too widespread and visible to be interpreted as mere abnormalities in a 'normal' development process. Experience has shown that in all industrialised countries the development of small companies does not follow a dynamic dictated purely by cost and trading mechanisms, but follows a precise territorial logic. The foundation of this territorial logic is found, as is well known, in the Marshallian concept of *local external economies* (or location economies). In addition to the economies of production that a plant derives from its own resources (from internal organisation and from management efficiency, and which thus fall under its direct control), a second stream of economies can be recognised that depends on the social relations of production that form outside the plant, but within the territory where it is located. The formation of significant 'bunches' of small companies occurs, in other words, in close association with limited socio-economic contexts characterised by historical conditions that 'explain' and 'describe' a social structure, a labour market, and technical-production interactions between local actors.

The logic of the industrial system thus becomes more complex and varied. First, added to the development of the small company as the expression of processes of externalisation and disintegration, are processes that form and develop production systems with clear territorial configurations. Second, the emergence of these *local systems* has brought political, social and institutional questions to the fore. These changes have brought three innovations in theory: (1) acknowledgement of alternative development paths and solutions to achieve economic efficiency; (2) acceptance of the inadequacy of purely economic explanations of local economic development; and (3) recognition of the role of social dynamics and institutional structures in the shaping of local systems (Coleman 1988, Goldenberg and Haines 1992, Putnam 1993a, Tarrow 1996). The debate on industrial districts is now based solidly on research concerning the *social characteristics* of territorial production systems and on *civicness* as a fundamental ingredient of development and modernisation.

Systems, networks and evolutive relations

In both the transaction cost approach and the theory of Marshallian external economies, the social processes that explain the dynamics of economic actors cannot be separated from those that give meaning to the idea of *local development*. This is so because it is locally (territorially) that human actors find most of the reasons for their own actions. Geographical proximity is vital to explain 'new industrial spaces' in terms of transaction costs, and a historically and geographically specific structure is the basis of the approach in Marshallian terms. It is important, however, not to lose sight of the distinction between the two theoretical proposals. The first is an atomistic approach, centred on individual behaviour (or on numerous types of individual behaviour). The second is a holistic approach in the sense that the system - the set of actors and relations between the actors - is seen as a whole. In other words, the company as an actor with its own strategic behaviour tends to be annulled, making space for *a new hybrid form*, the district company, which transcends individual behaviour. This distinction gives the sense of the real terms of the debate between the supporters of the two positions. Both identify advantages and disadvantages that can be evaluated differently according to their points of view. The former asserts a causal dynamic with general validity, while for the latter it is instead essential to go beyond the limits posed by the 'great' theoretical constructions, countering them with a 'weak', pluralistic explanation that recognises the variety and complexity of forms in which contemporary reality manifests itself. Nevertheless, both leave numerous problems unresolved.

The *network* approach embraces the entire local system without losing sight of the relations between individual actors. By interacting reciprocally, actors are the constituent elements of the system. Thus, local dynamism does not result from companies' individual actions, but of their collective behaviour. In this way, the assumptions of the Marshallian approach are not denied, but they are freed within model of peripheral Italy. Without going into detail, it is enough to recall the importance of the theoretical proposals of the Italian model. On the one hand, the model helps to define a dynamic approach to company development, in that its foundations lie in a complex set of *relations* (local and other) that confer a *strategic-operational identity*, i.e. a capacity to relate to the external environment. On the other hand, there is a shift from the micro level of analysis (the individual company) to the intermediate level, and so the company becomes an *evolutive* actor (or, better, *co-evolutive*) in relation to an environment which constrains it or presents it with opportunities.

The *evolutive* dimension of the company (and at the same time of the local system) thus explains economic and extra-economic issues together, expressed in terms of both *collaborative* and *competitive* relations. These relations of *socialisation* (Giddens 1984) cannot be represented in purely functional terms as trade relations involving transport and accessibility costs, but need to be interpreted from the perspective of the organisation of the system as a dynamic structuring process. It is for this purpose that the notion of *network* was introduced (Thorelli 1986, Butera 1990, Holton 1992, Håkansson and Snehota 1995). The explanatory context is thus explicitly *systemic*. The identity of the individual actor is not annulled in the network, but *co-evolves* with it and the actors that comprise it. The set of network relations determines the organisation and autonomy of individual firms in relation to the external environment. The character of the system is not therefore dependent on the 'general laws' of the economy, but on processes of local organisation with their own rules.

Just a few references are enough to give an idea of the effects of network organisation on the actors that constitute it. First, the exchange of information (financial, fiscal commercial, scientific and technological) between companies is achieved both within the framework of market rules and outside them. It is based on trust, planning for the future, and on more or less tacit norms accepted by the members of the system. This set of norms also facilitates the social control of information entering from the outside, adapting it and making it consistent with the local socio-economic context. In addition, this encourages the development of a certain collective identity, grounded in a dynamic of co-operation and competition between actors. Second, there is relatively systematic, formal or informal, agreement between companies and institutions that eases the exchange of information on technology, business and competition. This consensus involves different forms of horizontal and vertical co-operation and facilitates the creation of a complex set of formal and informal transactions. Third, the development of a technical culture multiplies the number of actors oriented towards technological and organisational innovation. In other words, the structuring of network relations eases the sharing of information and the generation of new organisational forms. The market only transmits, in fact, a small quantity of information, and even when this happens it can be at high cost.

The network dynamic is, therefore, the basis of a process of *collective learning* that stimulates the evolution of the actors operating in the system and regulates their relations with the outside world. It is in this way that it has made a contribution to explaining economic systems in terms of *relations between the local and global*. The local and the global are inseparable and yet they each pose their own questions. If the small company operates on world markets *indirectly*, access to globalisation occurs through their inclusion into international networks autonomously or in connection with other internationalised companies. Access to networks allows companies (which remain locally orientated) to operate on global markets indirectly, both to receive resources (from other members of the network) and to supply their products to the market (Julien 1995).

Seen in this light, territorialised SMEs' access to globalisation makes the local system (the industrial district, for instance) dynamic. A distinction must be drawn, however, between enterprises aiming at internationalisation more or less explicitly and others that draw advantage anyway from the new global position reached by other actors. All this is very important in debunking the paradigmatic vision of the district in which the enterprise as actor tends to disappear to be replaced by a new hybrid form, the *district-enterprise*, which behaves not too differently to a large corporation. This scenario re-introduces very forcefully the enterprise's role as actor and leads to a less deterministic vision of the district itself. As globalisation advances, the integrating forces within the district lose bite and some dynamic actors, opening up to the outside, break the correspondence between the district and its operators.

The learning economy

Useful as the network and systems concepts are, they leave significant aspects of local development unresolved. The significance of the concept of *social interaction*, essential to a network dynamic, has not so far been specified. If we stopped at this point, we would in reality be failing to explain a chapter in the evolution of social science thinking. A significant step forward in thinking came from the interplay between a range of disciplines including economics, sociology, anthropology, philosophy and the territorial sciences (Geertz 1983, Granovetter 1985, Johannisson 1990, Lundvall 1992, Lundvall and Johnson 1994, Maskell et al 1998).

The network is a representation of a set of relations of interdependency between actors, and the dynamic of the system is brought out as a *learning process* dependent on the variety, quality and density of information. At the same time, information is by its very nature a complex resource that can rarely be produced by a single player - it is 'in the air' of a place, as Marshall observed at the beginning of the century. Sharing and co-operation are thus required to produce it, i.e. interaction and communication. But what is the mechanism that generates the evolution of the system through the sharing of information? How does social interaction between the actors involved occur? Also how can we explain the greater dynamism of certain systems compared to others less able to produce, control, and use information for development purposes?

The fact that the cognitive dimension has established itself recently in the re-founding of economics and management science is well known, but this is only of marginal concern. What is instead essential is that these concepts show very strongly that a production system cannot exist independently of the territorial context within which it co-evolves. Again views on the nature of the small firm and locational dynamics have been essential in overcoming the limitations of conventional economic theory.

Nonaka's contributions have been important in this context (Nonaka and Takeuchi, 1995). He re-introduced the thesis, already put forward by Polanyi (1967), of the dual nature of knowledge; of *codified knowledge* and *contextual* (or tacit) *knowledge*. The latter, in particular, is an expression of the specific socio-cultural environment that expresses it.

It is rooted in the actions of actors and is valid only in the environment that created it (though codified knowledge can also be accumulated locally). This new way of looking at reality is based on the idea that a system of production cannot be founded on only one type of knowledge, but on the reciprocal integration of both types of knowledge. If this is true, it then follows that the local dimension is a key element in the evolution of the system, since it is locally that fundamental knowledge conversion processes occur (Becattini and Rullain, 1996, Asheim and Isaksen, 1997). The local system maintains, in fact, a dual function. On the one hand, it creates the link between codified knowledge (explicit and transferable) and production activities. In this process of conversion, contextual knowledge, as an expression of values, culture and well-rooted organisation allows the local system to filter and transfer codified knowledge (of global origin), adapting it to its own needs. On the other hand, it converts contextual knowledge into explicit knowledge. A system's competitive advantage derives from its capacity to transfer knowledge produced locally and to relate it to the global circuits of exchange of codified knowledge.

A local system, understood as a place of integration between contextual knowledge and codified knowledge, is not, therefore, a closed system, but a segment of a (virtually global) circuit of learning and production of new knowledge. This does not mean that the world can be represented as a mosaic of different local systems, all equally competitive and capable of valorising their own accumulated knowledge. Local systems with the greatest capacities for production and innovation will be those in which there is continuous and more intense interaction between the two spheres of knowledge, i.e. those capable of activating and enhancing their own distinctive substratum of values, know-how and institutions. A local system re-appears, in this light, as an organised system endowed with a specific identity. It is on the basis of this identity, essentially, that the system evolves, responding to external stimuli. It is the foundation for the dialectic on which the contemporary economy is based.

The fields of communication

The discussion of the previous section reveals two important elements for understanding the dynamics of local economies. First, there is the recognition that an interactive process aimed at creating and developing knowledge does not depend on simple transactions regulated by formal relations. Second, to understand these relations, the metaphor of the network has been introduced, used to represent the relations between actors that transcend competitive conflict and enhance communication. If the argument was based purely on generic resources (such as raw materials, services, manpower etc.) the locational behaviour of economic actors could easily be explained in terms of cost differentials. The location problem would thus appear as one aspect among many that contribute to the definition of strategic behaviour. The concept of a *field of communication* represents instead *specific resources*, which, being explicitly localised, make territory itself a strategic resource in the process of change. Change then ceases to be a mere problem of the introduction into a place of new knowledge developed outside the field of action of a particular set of economic actors. 'Place' and the local environment are pivotal in the creation of knowledge. Very briefly, the environmental (territorial) conditions on which a learning process is built are:

- informal and non-trade relations between economic actors;
- a shared technical and industrial culture;
- historically consolidated collective behaviour and practice; and
- an entrepreneurial and technological atmosphere.

Clearly, the communicative dimension of these conditions (the solid foundations of all the conditions listed) discriminates between a set of specific resources and a set of generic resources. In contrast to the generic, specific resources are explicitly localised. It is unthinkable, in fact, to imagine that, being produced by a given context through the historical evolution of relations between actors, they could be reproduced in a different geographical area to their original one. As is well known, this set of territorially non-reproducible conditions, expressed by the term *milieu*, means a system of actors and structures that can be fully grasped only within the complex play of reciprocal interaction.

The approach is explicitly holistic. The set of interactions between the actors and territorial conditions present is the origin of a system effect that shows itself in a particular technical, political and social atmosphere, climate and culture. It is thus a process that reproduces its own coherence as it evolves. Given these premises, it is logically difficult to reach a full and exhaustive definition of a concept - the milieu - that can be grasped only through many qualitative dimensions (Camagni 1995, Maillat et al 1991). The examination here is limited to two essential aspects of the concept in question. On the one hand, there is its dynamic character, given by the complex play of interactions activated in a particular environmental context. On the other hand, is the fact that this set of specific resources does not constitute a simple condition of cost reduction (that economic actors can, at best, find in many places), but a set of non-reproducible externalities (economic, social, cultural and environmental) that have accumulated through long-term historical processes. In other words, other criteria are added to the efficiency criteria of conventional economic analysis (obviously not cancelling them), giving economic action a contingent character, stemming from its 'embeddedness' within a given cultural, political and social context. This explains why a learning process is necessarily localised. As already noted, a network dynamic cannot be separated from the environment that plays host to the relations between actors. On this point, Lecoq (1993, pp.95-96) offers a convincing and synthesising interpretation of the concept of the milieu, as a *milieu innovateur*. Its features comprise four interlocking dimensions:

- a *territorial dimension*. The *milieu* is a geographical space whose boundaries cannot be defined *a priori*, but depend on the identification of the specific behaviour that gives it coherence;
- an *organisational dimension*. It is the co-operation between the actors, open to each other, that is the origin of the production of innovation. This organisational logic takes shape in a complex system of interdependence and reciprocity that usually assumes non-commercial forms;
- a *learning dynamic*. In the play of interaction, the actors modify their behaviour to produce new production and organisational combinations. Learning is collective, therefore, in that it is based on knowledge shared by various partners; and
- an *industrial culture*, that expresses the milieu's historical memory, knowledge and technical background. This finds concrete expression in professional practices, a work ethic and shared values. It is the source of a certain consistency of behaviour and, in change, ensures the stability of the system.

From this point of view, learning is substantially a socialised and collective process (as well as an economic one) based on a territorialised organisation of relations between actors. It is organisation, in other words, that defines different paths for the creation of different knowledges, as the repositories of non-reproducible economic, social and institutional practices. In this light, a system is open to information from external sources and is, at the same time, the repository of specific externalities, which are organised, co-ordinated and related to the economic, cultural and technical structure. These structures ensure that local (tacit) knowledge is valorised in new technological and production solutions.

The local synthesis: a geographical revenge

The brief summary outlined here of some of the current lines of thought on processes of economic change is inevitably partial. Rather than examining the details of one or other theoretical proposal, the goal has been more limited - to focus on broader theoretical arguments that offer different perspectives on the issue of local economic change. From the discussion six broad points emerge.

First, it is evident that the perspectives reviewed here converge by highlighting *the local system as an intermediate element of analysis*. The territorial dimension emerges from the need to go beyond the contrast between the macro-economic level of the national systems (the globalisation process weakens the economic sovereignty of the nation states) and the micro-economic level of the company-actor. The local system is first and foremost the level at which a *relational dynamic* occurs between different actors. This confers on both the actors and the system itself the capacity to assimilate the disturbances and changes that arrive unceasingly in the technological and competitive world, as well as collectively producing innovation and knowledge. As we have seen, this places the dynamic of the actors in their historical, social and institutional setting, thus *contextualising* the global evolution of the contemporary economy.

Second, the wide-ranging debate on local systems has helped to undermine much of the dogma on which conventional economic theory is founded. At this point in the debate, the criteria for the definition of these new production entities are no longer an issue, just as whether they represent or not the superseding of (or a variation on) Fordism is not an issue. The debate has led to the specification of a set of theoretical and methodological instruments that have a more general value, and are thus applicable to the many diverse forms of the contemporary economy, from districts of small companies to old centres of mass Ford-Taylorist production.

The debate could not have developed without overcoming the positivistic approach to social science, which led to analyses of economic phenomena by abstracting them from their cultural, historical, social and institutional settings. Assuming social and cultural context to be the key variables in the organisation of production allows us to lay bare the theoretical trap of orthodox economics. Orthodox, economics bases explanation purely on *strong* conceptual (or technical-mercantile) categories, making it incapable of explaining the importance of changed market, technological and non-trade conditions on the relations between actors (Grabher, 1993a). In geography, this approach characterised the functional-organisational approach to understanding the Fordist company, centred on processes of technical and economic concentration and the 'virtuous' interweaving of companies, parts of companies and *space*. This form of reasoning was fundamentally suspicious of so-called *weak* conceptual categories (such as 'identity', 'industrial atmosphere', 'communicative interactions', and 'industrial culture') which, though difficult to measure, have become more prominent owing to growing economic interdependence and the breakdown of spatial and temporal barriers.

Third, these weak conceptual categories are the origin of the thesis on *territorial company embeddedness*, introduced by Granovetter, a sociologist, and then discussed in spatial terms by, among others, Powell (1990), Granovetter and Swedberg (1985) and Taylor (1995). In evident contrast to the orthodox economic explanations, the reference is no longer to the company as the organisation governing the economy, but the *formative processes* of the organisations themselves, which derive from collective behaviour (networks, in essence) expressed both inside and outside the market context. These processes are instead the expression of an *embeddedness* which is cognitive, political and cultural (Taylor 1995). The network organisation that characterises the contemporary economic system is increasingly based on technical, organisational and communicative interactions. These *non-material, intangible economies of location* are difficult to transfer from one place to another precisely because they are of an intangible nature and specific to each context. It is on these intangible economies that the organisational logic of the contemporary economy is based.

Fourth, this way of conceiving the interaction between the company and its local environment leads to a redefinition of the problem of the competitive dynamic both of *large* and *small* firms, so that the contrast between the two, while maintaining fundamental, specific features, seems in fact to lose theoretical importance. The 'dual convergence' of these two company structures had already been understood by Piore and Sabel in their analysis of the new importance of the regional economies in the era of flexible specialisation. More concerned with the technological causes of new production phenomena than with the territorial conditions in which they develop, the two American economists still proposed a 'transgressive' point of view which could not be acknowledged as representative of a real dynamic. Starting from the theses of environmental embeddedness, this perspective now finds a more mature position.

It is not, in fact, enough to state that the large oligopolistic corporation has begun to implement flexible organisational forms, once the exclusive dominion of small and medium sized enterprises. This may be true in some sectors and not in others, in certain national realities and not in others. Assuming instead that competitive advantage derives not only from interaction with other local companies, but through embeddedness in a specific socio-economic and institutional context, this thesis becomes common, within certain limits, to both small and large firms. Growing environmental complexity induces small production units as much as global companies to search for new strategies of behaviour aimed at constructing more effective relations both with other companies and their socio-cultural contexts of reference and with places (local systems). In global competition, in fact, *the diversity of strategic solutions becomes the foundation of competitive advantage* and the mechanism that produces economic value.

Fifth, in this framework, the *co-evolution of enterprise and environment* - understood as a specific set of tangible and intangible conditions - appears to be a pre-condition for company development and, at the same time, a factor in the reproduction of the identity and diversity of the local systems. Just as the socio-cultural environment is not merely an expression of autonomous historical evolution, but is also dependent upon the strategic behaviour of economic actors, the company can no longer be thought of as a system which is economically and organisationally self-sufficient and which establishes hegemonic relations with the environment. On the contrary, it appears now as just one of the actors between which a complex dialectic relationship is constructed.

Sixth, this set of issues opens the problem of regional competitiveness in an era of globalisation. It goes without saying that the chance of a regional economy launching itself successfully onto international markets lies, on the one hand, on the products it sells and, on the other hand, on the fact that competitive advantage is no longer necessarily found in the exogenous search for the best technology and production methods available. Increasingly, competitive strength has to be sought inside the region itself, in the capacity for co-ordination between producers, consumers, institutions and other local actors. Above all, in industrialised countries with high production costs, the problem of competitiveness depends increasingly on the capacity to create, accumulate and utilise knowledge more rapidly than competitors (Maskell et al 1998, Chapter 2). This is accompanied by growing international product specialisation (a phenomenon that, at first sight, is surprising in an era when the use of communications and computers encourages as never before the diffusion and imitation of technology) (Salais and Storper 1993). This means that the growing specialisation of national and regional economies is no longer dependent on economies of scale in production. Instead, it is dependent on the nature of the products put on the market, on the know-how to make those products, on the nature of the needs they satisfy, and on the continuously evolving capacity to make products and preserve their originality. Local economic growth comes from communication, accompanied by the implementation of *network strategies* - between companies, between companies and institutions, and between different institutions - where the creation of *social capital* regularly supports the formation of small enterprises (Cooke 1995, Indergaard 1996, Morgan 1997, Rosenfeld 1992).

The two faces of Janus

The question of the small company thus has great theoretical, epistemological and methodological importance, which goes beyond the specific phenomenon of SMEs. In this framework, *territory* is the central concept capable of giving intelligibility to the complex turn-of-the-century economic dynamic, while providing a synthesising, interdisciplinary dialogue that conflicts with the old categories of thought. The outcome of twenty years of reflection is, in this light, significant: the thesis is that geography returns to being a *science of places*, not in the banal meaning in which rationalism had confined it, but as the expression of pluralistic knowledge.

To reach a conclusion, it is worth going back to the real phenomena from which the discussion in this chapter started. If the small enterprise appears today as a success story, or at least particularly interesting from the point of view of the historical evolution of economy and society, it is not because it represents a more human and more ecological alternative to mass production (repeated as a naïve mantra by the supporters of *'small is beautiful'*). It is because recent developments since the 1970s have gradually undermined the reasons that supported the division of labour between large and small companies, set in the Fordist-Taylorist paradigm.

The first to suffer from the decline of Fordism were, as is well known, the large companies. The small company, in contrast, has appeared particularly vivacious not because it is endowed with some intrinsic competitive superiority, but because it has a greater ability to adapt in the short term to the changes that have hit the competitive advantages of large companies. It is clear that all that glitters is not gold. Many small companies have kept the characteristics of marginality, including a certain indulgence towards the black economy, the use of low-skilled labour, weak financial structures and too informal an organisation. However, the position that it has come to occupy in economic thought and in the evolution of the real world has nothing to do with a hypothetical weighing up of the positive and negative aspects of small enterprise. It is not a question of discussing whether the small company is beautiful or ugly: the problem is that the passage from Fordism-Taylorism to post-Fordism has left a certain vacuum in the reference models on the role and activities of small companies.

The small company is not at the heart of academic and political debate because of its size and traditional competitive weakness but because of its identity and its socio-cultural qualities. The qualities reflect the situation that:

- the small company should not be considered, in itself, as a core business compared to other businesses, but as the framework of an entrepreneurial society based on the principle of *risk sharing*;
- the small company cannot be considered in isolation. It has become a model relevant at the scale of global systems, involving both *conflict and co-operation*; and
- in a risk sharing society, political power needs more direct contact with business, needs to shift industry policy to the local level, and needs to assume the costs of research and infrastructure.

Relations between companies are no longer seen as purely competitive, but as involving complexly interwoven competition and co-operation. The irreconcilability of these two concepts is more apparent than real. The relationship between competition and co-operation had been lost in the half century of Fordism, but it is central to the Japanese model of industrial organisation discussed in the next chapter. This relationship is seen in the supply chains of the Japanese corporate giants and in the relations between large competing corporations, discreetly arbitrated by the MITI. There is also the industrial district model, both Italian (where the cultural and social 'cement' is essential) and of a more general technological type (in which sharing reflects a technological imperative). The risk sharing society is not thus, in itself, a society of small enterprise or of closed local systems, but a place of variety. The contemporary economy is an economy based on differences, on the exploration of many different paths of evolution, leaving history to decide on their validity. In this sense, the small company ceases to be a special case, and the 'local' becomes the foundation of an economy of diversity and communication between the different actors. This is, in the end, the essence of the post-Fordist economy.

The analysis presented here started with the first studies of industrial dualism. Those first studies were fully consistent with a system of Fordist-Taylorist relations, dominated by hierarchies (both inter-firm and intra-firm) and competitive relations. Even then, however, the district economies were an exception, where the division of labour between a multitude of small companies introduced elements of co-operation between highly competitive players. But these districts were more or less marginal phenomena, whose production was for niche or local markets.

The Fordist equilibrium was overturned by: (1) the arrival on the scene of newcomers (Japanese corporations, small Italian producers, American high-tech companies); (2) growing internationalisation (with its high costs and the subsequent search for ideal partners) and, (3) the closer relationship between science and production which put companies in a position where they could not realistically dominate the advance of the technological frontier. These changes raise fundamental questions about the need for inter-firm co-operation. They put large companies into the front line and needing to emulate the practices of small firms (seeking flexibility, product personalisation, and embeddedness in local markets). At a theoretical level these changes have two fundamental features. First, they reconfigured the operational capacity of large corporations, which shifted into market sectors traditionally reserved to small firms, excluding many small companies from international co-operation, which occurs between the leaders in the major sectors. Second, they forced small companies to learn new and different forms of working, new approaches to technology, and new product and market strategies.

These are demanding transformations for small companies. Curiously, the co-operative 'model' on which the rise of the small company was based hides many pitfalls for them. The pitfalls can be better understood by recognising four different types of small company; three types of *modern* small business, confronted from birth by non-local markets, and a fourth *traditional* form of business, operating in almost exclusively local markets.

The first form of modern small business is the *district company*. Part of the changes that have been seen in large companies can also be seen in the districts. Faced with the challenges of competition, there was downsizing, closure, acquisition, and merger that modified the internal geography and relations with the outside. As a result, the district value chain opened up as is illustrated in Chapters 11 and 12. Companies downstream could no longer be content with the district's internal, captive demand, and went looking for new markets. In their turn, the downstream companies were no longer forced to produce inside the district, and could move production to other markets. It is obvious that this is possible only for the larger companies, the leaders, which could open up channels of communication with selected partners.

The second type of small business is the company operating in *intermediate markets*. For them, the essential question is change in their relations with large firms which, in the Fordist era, were based on criteria of cost and/or loyalty. This type of relationship showed all its failings when large companies began to assess their suppliers on the basis of their competency, their innovative behaviour, and their ability to guarantee quality. Relationships between customers and suppliers now became a mixture of competition and co-operation, as analysed in Chapters 4 and 5. For this type of company, the development of internal technological intelligence is essential to evaluate shifts in the technological environment and to be able to co-design the continuous development of products and processes with larger company counterparts.

The third type of modern small business is the small company that *serves its own markets and is not linked to large corporations*. This type of company is in a phase of profound transformation. Internationalisation, technological evolution and market segmentation have made skills available to it that were previously reserved only for large companies. But, at the same time, this type of firm requires 'filters' to gain access to technology, market information that can only be found externally.

The last type is the small company that operates in *local consumer markets*. It is above all this type of company that finds itself in serious difficulties, as the pressure gradually increases from larger scale and specialised manufacturers. If we look closely, it is this type of company that faces the greatest problems from the crisis in Fordism.

It would seem unlikely that the behavioural solutions to changing conditions can be pursued autonomously by small companies, counting on their own self-sufficiency. It is necessary first to pursue the synergy between different players that only *network strategies* can offer. Second, an industrial policy is needed that supports the innovative transformation of companies as analysed in Chapter 8.

Conclusion: the network era

The new network era presents companies with opportunities to pursue a multitude of goals through network strategies. Network goals might include: a) establishing co-operative relations with one or more competitors; b) exploiting complementarities and specialising within the value chain; c) loosening dependence on large customers; d) organising relations inside the system and open relations with companies outside it; e) broadening traditional market niches to a territory defined on the basis of competencies; f) extending the number of countries in which the company operates; and g) creating a critical mass in order to enter new markets and set up manufacturing and service processes.

What brings these many goals together is the pursuit of *economies of scale in the use of knowledge* (technological and market information) which is not only produced together by its members, but intercepted by the network as a whole. These economies of scale in the use of knowledge do not arise from a contract that obliges the parties to act in unison, but from the satisfactory outcomes of a continuous *division of labour* which can be improved only through co-operation. However, if types of interaction can differ so too can the types of network in which they occur, but all aimed at the development of a communications and interaction infrastructure that increases the effectiveness of the division of labour.

Once again, if an ideal general network model has to be prefigured, the most coherent, both theoretically and in relation to the current phase of transition to post-Fordism, is represented by the network understood as a *self-organising system*. The concept of self-organisation has two fundamental meanings. The first meaning expresses the capacity to *transform disorder into order* in other words to learn through exploration, verification and interpretation that the network allows a number of companies to act together. In this light, the network is a *learning system* - not designed to implement a pre-determined plan, but to learn from the emerging circumstances. What makes the network capable of learning is the fact that many decentralised intelligences have access to all the information pooled in the network. The second meaning expresses the possibility of *creating order from disorder* on the basis of action plans formulated by one company and shared by others. A network is made up of 'intelligent (and active) nodes' which, although communicating with each other can obtain the contractual agreement of others.

The problem remains of how to create appropriate policy – a *network policy*. The local environment and the institutions that are expressions of that environment have a fundamental role in stimulating network formation. It is in the local environment that enterprise puts down its roots. The problem is that small companies feed off their own environment. They now need to learn to operate as part of a collective enterprise, and to develop the local business community. In the industrial districts, it has been history that has determined this co-evolution. In other cases this culture has to be created. In post-Fordist society the challenge for small companies is coming to grips with territorial risk sharing.

3 Industrial Network Formation and Regulation: The Case of Japan's SME Policy[1]

Yuko Aoyama

Introduction

The purpose of this chapter is to better understand the relationship between policy and the formation of industrial networks. There are potentially many programmes and policies from various arenas that may facilitate the formation of industrial networks. Local economic development and revitalisation programmes, employment creation programmes, and high-tech promotion programmes all potentially affect the formation of industrial networks. This chapter focuses specifically on SME policy and its role in promoting inter-firm collaboration. SME policy is one of the more prominent and direct means through which governments exercise influence over industrial organisation, by supporting the survival of smaller firms in the market. In Japan, SME policy has been used to help structure industrial organisation, to promote economic benefits and sustainable business performance in both large and small firms.

Japan's SMEs from a comparative perspective

Japan's SMEs survive by adopting two strategies. First, they rely on a long-standing craftsmanship tradition to generate incremental process innovation, and second, they have successfully shifted their operations from domestically oriented products to export-oriented products (Itoh 1958). Because of their deeply rooted craftsman tradition, many SMEs possess extremely sophisticated technical skills, far superior to those that their size would imply. Also, in contrast to managers within large firms whose job security is guaranteed, heads of SMEs are entrepreneurs who are highly motivated to produce superior products to ensure their firm's survival. Though sub-contracting ties to large corporations, the superior products produced by SMEs help these large parent firms to virtually eliminate all defects in finished products, thereby improving product quality. It has been argued that Toyota's lean production, for instance, would not have been functionally effective without the array of extremely reliable parts that Toyota purchases from its sub-contractors (Womack et al 1990). Thus, Japan's large corporations clearly benefit from collaborating, controlling, and taking advantage of the large pool of small firms. SMEs bring flexibility into the overall production structure, generating benefits for both small and large firms. Thus, although much of the control and power may reside at the top of the network hierarchy, the figures on SMEs in Japan clearly indicate that Japan's SMEs occupy an important role in the economy.

Table 3.1   Definition of SMEs in Japan according to the Small and Medium Enterprise Basic Law

Sector	Capital (1989 prices)	No. of Employees
Mining, Manufacturing Transportation, Construction	Y100 million or less	300 or less
Wholesale	Y30 million or less	100 or less
Retail, Services	Y10 million or less	50 or less

Source: MITI 1989

In Japan, SMEs are usually defined as firms with fewer than 300 employees. However, more detailed definitions exist, in terms of numbers of employees and the size of capital (see Table 3.1). According to MITI's definitions, SMEs accounted for 52 per cent of sales of manufacturing products, 57 per cent of manufacturing value-added, 62 per cent of wholesale trade sales, and 78 per cent of retail sales (SMEA 1998). Furthermore, SMEs in Japan have constantly been the major source of employment throughout the post-war era. As shown in Table 3.2, Japan's SMEs have employed around 80 per cent of the total labour force over much of the post-war period.

Table 3.2   Share of SMEs in total private sector employment in Japan, by sector

Sector	1954	1972	1977	1981	1986	1991	1994
	Per Cent of Total Sector Employment						
Mining	32	64	70	76	76	85	88
Construction	85	90	94	95	96	95	95
Manufacturing	74	69	74	74	74	74	67
Wholesale, Retail trade	98	86	88	87	87	86	84
Banking, Insurance	90	83	84	86	87	85	86
Real Estate	95	96	98	98	98	97	97
Transport, Communication	--	84	87	89	88	87	88
Utilities	--	66	67	67	69	74	73
Services	98	72	71	70	67	64	63
All Sectors	84	78	81	81	81	79	77

Note: The definition of SMEs follows the categories established by MITI, as shown in Table 3.1. 1954 figures also include public organisations.

Source: Sorifu, jigyo sho tokei chosa hokoku sho

Table 3.3 shows cross-national comparisons of the share of SME employment for G7 countries. The share of SMEs jobs in total employment in Japan is the highest among the industrialised economies, and is particularly prominent when contrasted with the same figure for the United States, which has fluctuated around 50 per cent since the 1950s[2] (Aoyama and Teitz 1996, Harrison 1994a, OECD 1996a). The contrast between the two countries in terms of manufacturing employment is even more striking. While more than 70 per cent of Japanese manufacturing employment is currently in SMEs, in US manufacturing the figure is less than 40 per cent.

Table 3.3 **G7 country comparisons of SME shares in employment**

	Employment share	Date
Canada	60	1989
France	69	1990
Germany	66	1988
Italy	49	1988
Japan	74	1991
United Kingdom	67	1991
United States	54	1991

Source: OECD 1996

The data pose two questions. Why are SMEs so much more important in Japan than in other industrial economies, and why is the structure of industry in Japan's manufacturing sector so significantly different? To answer these questions, the next section explores the various ways in which Japanese SME policy has intervened in the economy throughout much of the post-war period of industrialisation. The frequency of state intervention in the Japanese economy is well known. However, most attention has been paid to the relationship between government and large conglomerates, and relatively little is known outside Japan of the role of government policy that specifically targets small manufacturing firms. The history of policy development in this arena reveals that since the pre-war era Japan's SME policy has consciously and explicitly promoted inter-firm collaboration at various levels. This extended history of policy involvement shows that policy can significantly influence and shape industrial organisation and encourage industrial network formation, given sufficient time and the appropriate business and policy environment.

Institutional paradigms and the role of SMEs

With the microelectronic revolution and the greater potential to build networked forms of industrial organisation, academics and policy makers have renewed their interest in SMEs as primary agents of industrial network formation.[3] As parts of networks rather than acting individually, SMEs are thought to exert a greater positive influence on the overall economy, and industrial networks themselves are seen as an alternative form of industrial organisation to large, mass-production based structures.

The recognition of this alternative has sparked theoretical debate over the possibility of capitalism having multiple and diverging development paths rather than following a single, linear model.

As Conti has outlined in the preceding chapter, SMEs have long been viewed as an inherently inefficient form of industrialisation, representing a form of capitalism that can be labelled as archaic and pre-modern. Traditional economic theories of the left and the right, such as those represented in Rostow's stage theory and traditional Marxist theories,[4] did indeed see SMEs playing their greatest role in less developed forms of economic growth. Such a linear view of the role of SMEs in an implicitly universal developmental path has been widespread among economists and policy makers in the United States as well as in Europe in much of the post-war period.[5] This view was particularly prominent during the heyday of mass production. Thus, the co-existence of large and small firms in the economy, the characteristic which Conti refers to as the "duality of industrial systems of the 20th Century", was perplexing to theorists. Many concluded that SMEs would eventually be phased out and be dominated by a more advanced, large corporation based industrial system.

The view that capitalist development can occur through varying and diverging paths recognises the different patterns of social institutions, cultural norms and values that affect economic organisation. Not only has the French Regulation School contributed significantly in reviving the institutional paradigms proposed earlier by Polyani (1957) and others, but it has also highlighted the possibility of divergence in capitalist development. In addition, the Regulation School was one of the firsts to identifying the *positive* role of economic and social institutions in the market (instead of the widespread neo-classical assumption that institutions typically function as *barriers* in the market). The Regulation School greatly influenced the work of Piore and Sabel (1984) which led to the introduction of 'flexibility' as an essential component of competitive and sustainable industrial forms. These ideas have been explored in numerous empirical studies that have adopted a range of approaches as discussed in Chapter 2.

While scholars generally agree that institutional and cultural factors play a role in developing a specific form of industrial organisation, opinions are divided as to whether government policies can function as a principal agent of change. Regulatory frameworks, which can vary from specific government policies to nationally implemented administrative structures and institutional environments, are considered to have a profound effect on economic organisation (Clark 1992a). The role of policy has been emphasised by the successes of the Asian economies up to the mid-1990s (see Amsden 1989, Gold 1989, Putnam 1993b, Evans 1995).

Christopherson (1993) has also argued that the regulatory framework remains an important factor shaping economic organisation in the US, despite the assumption of the declining power of state intervention in the face of globalisation. On the other hand, empirical studies on industrial districts in Silicon Valley as well as in Emilia Romagna make reference to supportive local or regional policy, but the role of government programmes are almost never more than the provision of guidelines. At the regional level, policies and programmes are seldom specifically targeted intentionally to increase inter-firm collaboration. While today policy makers scramble to find a way to induce successful industrial networks in their regional economies, much of the actual research on industrial networks is inconclusive on the role played by policies.

In contrast, the study reported here reveals that Japan's SME policy encouraged, supported, and provided an institutional framework for inter-firm collaboration throughout the post-war period. It is argued that policy interventions, in this case at the national level, can in fact help shape particular forms of industrial organisation, given certain political and economic contexts. The choice over what society considers the best mechanism for economic growth has a significant impact on the way inter-firm networks are formed in various societies. Since the Japanese industrial networks *per se* have been studied extensively, this chapter focuses on the policy side of industrial network formation. Japan's SME policy will be examined to evaluate how dominant ideologies, historical legacies and economic beliefs have affected policy formation and programme objectives.

Researchers both in Japan and elsewhere attribute much of Japan's post-war economic success to effective co-ordination, collaboration and mutual dependence between large corporations and SMEs, particularly in the manufacturing sector (Nakamura 1983, Miwa 1990, Dore 1986, Gerlach 1992). It has been argued that competition among SMEs for contracts offered by large corporations has contributed to improvements in quality for many of Japan's industrial products (Womack et al 1990). Although much of Japan's economic organisation has been portrayed as an economy driven by large and powerful conglomerates, now widely known as *keiretsu*, it was, thus, these numerous small companies that allowed large Japanese firms to function effectively (Friedman 1988).

In Japan, much entrepreneurship leading to the formation of SMEs in the industrial era originated in craftsmanship traditions (Whittaker 1997). As in Europe, Japan's SMEs arose from an artisan tradition, which began as agricultural surplus labour co-ordinated by a wholesaler - known as the 'tonya' system. In this system, production takes place within individual households, and merchandise is gathered by the wholesaler, who might return for more orders, sometimes with new or added specification. Eventually, the locality develops a set of specialised skills in a particular industry, and the foundations for an industrial district are put in place.

SME policy in Japan: tools for developing 'industrial systems'

The origin

The origin of SME policy in Japan, started seriously in early 1930s.[6] The Ministry of Commerce and Industry, the pre-war predecessor of the Ministry of International Trade and Industry (MITI) had begun to consider policies to promote sub-contracting arrangements between large and small firms before 1935. By 1937, the Ministry had initiated programmes that included assistance in financing, technical advice and management advice in 13 prefectures. These programmes aimed at promoting rapid industrialisation in the peripheral parts of Japan, with the purpose of integrating small firms into the national production system in order to mobilise economic activity for the wartime economy.

Policy makers were interested in the advantages that sub-contracting arrangements offered for the Japanese economy, and were already aware that through sub-contracting:

• the overall production capacity of the economy could be expands without large capital investments;
• economies of scale could be gained by small firms by producing for a number of firms; and
• improved product quality could be achieved, because specialised small firms had better skills and expertise than were available through mass production at the time (Toyoda 1941).

At the start of sub-contracting arrangements, however, relationships were initially quite chaotic. Some sub-contractors constantly shifted parent firms, and some parent firms refused to nurture, protect and share technologies with sub-contractors. The Japanese government responded by urging the industries to develop 'organic' relationships between parent and sub-contractor firms, advising firms to conduct businesses with a single parent firm which would in return be responsible for providing contracts, technology and training for employees. This was done in the context of the emergency wartime production scheme, in a period in which administrative interventions were highly effective in making private sector firms comply with government directives. With the rationale that firms should contribute to the development of an advanced defence-oriented state, firms were urged temporarily to drop their individual profit maximisation strategies for the sake of raising the overall productivity of Japan's industries (Toyoda 1941). In this way, Japan's current system of industrial organisation has been the product of a deliberate policies promulgated in that wartime era, which were implemented to achieve the specific objective of productivity maximisation in the face of limited resource availability.

Post-war policy during the occupation

Post-war policy toward SMEs began with programmes implemented by the General Headquarters of the Allied Forces, which occupied Japan in the immediate post-war period. In order to neutralise the political and economic powers that facilitated Japan's imperialism, the initial purpose of the Allied Forces was to dismantle the pre-war conglomerates, known as *zaibatsu*. The Small and Medium Enterprise Agency (SMEA), established under the jurisdiction of MITI in 1949, became a primary instrument in executing this policy and a mechanism for implementing a regulatory framework oriented toward the free-market model (JETRO 1989, MITI 1989a, b). As a result, policies promoting promotion of sub-contracting, inter-firm collaboration and sector-based industrialisation schemes were all dropped by the Allied Forces. The initial implementation of SME policy in Japan in the post-war period, therefore, was a complete reversal of the previous policy.

SMEA initially functioned as an agency whose primary objective was to save distressed SMEs by providing better accessibility to capital (Arita 1990, 1997). Several financial institutions were established to facilitate this policy, including the National Finance Corporation in 1949, created specifically to fund SMEs (JETRO 1989, MITI 1989a, b, Kurose 1997). Also, in 1951, a variety of regulations were established to assist private banks, such as trust banks and mutual banks, to make loans (SME Finance Corporation of Japan 1986).

The post-occupation policy

As the occupation came to an end, however, the initial post-war objectives of SME policies were altered (Chusho kigyo dan chusho kigyo kenkyujo 1987, Arita 1990). MITI announced target industry policy as the backbone of national economic policy in 1951. The primary objective of MITI's industrial policy was the improvement of Japan's international competitiveness. To achieve this objective, MITI opted for programmes that targeted mainly heavy industrial sectors, such as steel, coal mining, electric power generation and shipbuilding. Through this targeted policy, MITI aimed at achieving rapid industrialisation by increasing productivity and reducing the cost of raw materials, energy and transportation. In conjunction with these policies, MITI developed programmes that were designed to promote modernisation and technological advancement in these sectors (MITI 1989a, Arita 1990). There were three major strategies identified in these programmes:

- protectionism, with the strict control of trade, currency exchange and technology transfer;
- the provision of capital through various quasi-public bank channels (such as the Japan Import-Export Bank, the Japan Development Bank, and the Industrial Bank of Japan) and the provision of tax incentives for facility modernisation; and
- the loosening of antitrust regulations to facilitate the concentration of capital and close co-ordination among large corporations.

As part of the overall programmatic change, SME policy was transformed to better complement and support Japan's industrial policy. The main elements of SME policy shifted from relief-oriented measures to the promotion of sub-contracting arrangements between large and small firms with the primary purpose of raising overall productivity. The purpose of SME policy thus shifted from the promotion of free competition to the establishment of a complex industrial structure that could bring about rapid industrial growth.

MITI enacted SME policies that directly contradicted the initial post-war objective. The Law on the Organisation of SME Associations of 1957 explicitly and actively promoted cartel arrangements through the provision of tax and financial incentives (Arita 1990, Kurose 1997). Small firms were encouraged to form co-operatives, which played an active role in the exchange of market information within an industry, thereby increasing stability among desperate small firms during this period.

In addition, the National Federation of SME Associations was established (MITI 1989a). As an umbrella organisation for SMEs in all sectors, the Federation took an active role in encouraging closer relationships among firms through both formal and informal networks. As a result of these policies, a variety of industry co-operatives and associations were established. In general, these industry associations today exert strong control over many Japanese industries, and have considerable influence over Japan's economic planning.

Despite the major reversal in Japan's economic policies, the allied forces did not intervene to prevent this development. This was probably because of the changing political climate of East Asia, particularly the threat of communism in the region at the time. Japan's policy reversal coincided with the MacCarthy's Era of Red Purges in the US and the onset of the Korean War. Instead of dismantling the economic source of pre-war Japanese imperialism, the priority of the Allied Forces on Japan shifted to the promotion of rapid economic recovery to contain the threat of communism.

The foundation of modern SME policy

The basis for Japan's SME policy today was laid out in the Small and Medium Enterprise Basic Law of 1963 (MITI 1989a). This law established a framework for a variety of measures that had previously been enacted without co-ordination by consolidating all programmes applicable to SMEs. In this law, it was explicitly stated that the promotion of SMEs is a component of overall industrial policy, aimed at improving Japan's industrial competitiveness. The role of SMEs in the economy was defined as contributing to the improvement of the industrial structure and to achieving balanced national economic growth through the improvement of Japan's international industrial competitiveness.[7]

> "The targets set in this law, in short, are to promote the growth and development of small and medium enterprises in co-ordination with that of the national economy, and to contribute to the betterment of the economic and social status of persons working there." (MITI 1989a)

In addition to the programmes that facilitated collaboration among firms, the Law on the Promotion of Sub-contracting Small and Medium Enterprises was enacted in 1969. This law provided additional incentives for smaller firms to establish sub-contracting arrangements with larger corporations. Under this law, a comprehensive programme was developed which packaged financial assistance with training as well as consulting specifically over facility modernisation in every stage of production.

The changing international economic environment in the 1970s caused MITI to consider the necessity of yet another industrial restructuring in order to adjust to slower economic growth. An advisory committee established by MITI in 1979 urged that the Japanese economy should shift its basis from heavy industries to energy efficient, high technology industries (MITI 1989a). In response to this shift in national industrial policy, Japan's SME policy took on another feature, namely the promotion of business conversion. Although it has been considered largely taboo in Japan to implement a policy that encourages firms to change their line of business, two new laws were enacted in order to better facilitate business conversion among SMEs from relatively traditional, labour-intensive industries to new high-tech industries with more promising futures.[8]

SME policy in Japan distinguishes various sectoral activities, and tailors policies accordingly.[9] Policy promoted the restructuring of industrial organisation to raise production efficiency for the entire production system.[10] Facility modernisation and business conversions were actively promoted, not only to streamline the production process from small to large firms, but also to increase productivity and improve SME survival rates. It is evident that policy makers have been aware of sectoral differences and needs, and have a decided priority toward SMEs in certain sectors, especially manufacturing.

The emphasis on the manufacturing sector is also evident in the objectives of the Small and Medium Enterprises Modernisation Law, enacted simultaneously with the Basic Law in 1963, which specifies the industrial sectors to which financial, informational and training assistance would be channelled (MITI 1989a). The primary purpose of this law was to encourage the modernisation of facilities among small firms within the sectors identified as particularly important for Japan's economic growth. The Law was later criticised for its bias toward larger SMEs with lower investment risks. This bias was also an indication that the primary concern of the policy was not to provide assistance for disadvantaged SMEs, but rather to modernise firms so that they would become efficient sub-contractors to larger firms. The approach taken by SMEA to remedy this bias toward medium rather than small firms was to provide more sub-contracting opportunities for very small enterprises.

SME policy at the regional level

Local governments are actively involved in implementing, administering and physically housing SME policy related programmes and their staff members. However, policy formulation and programme initiatives typically come from the national government, and any changes in policy directions are determined at the national level. National authorities allocate funds, even if they are eventually distributed by local agencies. Upon receiving directions from the national government, local governments respond by designing locally based programmes which vary little between different prefectures. Local offices and facilities play an important role by providing easy access to services. For example, a number of technical centres known as *kohsetushi,* are scattered across Japan making technical assistance and management advice accessible to many small firms in the peripheral areas. In comparison to the US, where small business policy is actively used as part of local economic development programmes, Japan's SME policy remains very much a part of national level industrial strategy. Thus, the local economic development aspect of the policy is secondary to the primary purpose of improving industrial efficiency (Aoyama and Teitz 1996). As part of national industrial policy, Japan's small business policies have tended to encourage agglomerations that constitute highly efficient and highly concentrated industrial complexes. It was only when diseconomies of scale in these agglomerations began to generate inefficiency, that Japanese authorities became interested in spatial policies. MITI's Technopolis Program, which began in 1983 and was eventually terminated in 1998 after widespread failure, illustrates the first effort by the authority in primary charge of industrial policies to engage in explicitly spatial policies. But even then, MITI's policies are essentially spatial policy for the sake of industrial policy, and spatial considerations still remain secondary to Japan's small business policy. In other words, small business policy in Japan has not been actively used as a tool in spatial planning. Thus, efforts by Japanese planning authorities to decentralise economic activities from the major urban areas have so far not been effective, in part due to the heavily industrial nature of small business policies both at the national and regional levels.

Summary

In sum, Japan's primary orientation in SME generation in the post-war period has been that of sub-contractor development (Aoyama and Teitz, 1996). This has largely benefited Japan's giant firms, while subordinating the autonomy and financial independence of SMEs. Furthermore, SME policy functioned to strengthen Japan's industrial policy in much of the post-war period. As a scholar in SME theory contended, "SME policy in Japan is an industry systems policy", policies encouraged continuous restructuring for the purpose of improving international competitiveness.[11] Specific and sector-based measures were taken to encourage the modernisation of production processes, and facilitate SMEs to enter sub-contracting arrangements which, as a result, better integrated them with the rest of the economy and formed a tight nexus between large and small firms. The fact that the Small and Medium Enterprise Agency is within the jurisdiction of MITI reflects the strong orientation of the SMEA toward industrial development.

SME policy in transition: constraints on the industrial network approach and a move towards neo-liberalism?

Evidence suggests that the current (1990s) recession has transformed the nature of the relationships between large and small firms in Japan. Although small firms have traditionally been considered by large corporations to be a buffer against recession, there has also existed a degree of mutual reciprocity nurtured by long-term business relationships.[12] It has been reported that such mutual support between large and small firms may be wearing thin under the stress of globalisation. When the value of yen rose dramatically against the US dollar in 1986, 60 per cent of sub-contractors surveyed at the time responded that they would seek salvation from parent companies. A more recent survey shows that over 20 per cent of firms were at a loss in devising an appropriate response to the crisis; over 40 per cent sought to diversify their business relationships. Up to the mid 1990s, optimism was shown by numerous Japanese thinkers who claimed that Japan's SMEs possessed a strong technology based productive capacity which could not be displaced by foreign competitors.[13] Yet, despite the advantages of flexible, networked industrial organisation over mass production, events in the past few years have called such an assumption into question. In reality, Japanese industrial capitalism turns out to be far more hierarchical and far less flexible than perceived. As the recession continued, it put new stresses on traditional relationships.

As a result, policy makers have increasingly turned to entrepreneurship as the crucial source of dynamism to jump-start the economy back to its previously higher growth path. Two major policy changes have been devised, one is to restructure the industrial systems,[14] and the other is to encourage more independence and autonomy among SMEs, based on the Anglo-American model of free competition. On the one hand, this is a departure from most post-war policies, which promoted interdependence among firms. On the other hand, it reflects a growing trend among SMEs, with gradually decreasing reliance on keiretsu.[15] Japan's SME policy since the mid 1990s has moved toward entrepreneurship promotion programmes, encouraging free competition, innovation, and deregulation.

Other than being considered pre-capitalist, SMEs are most frequently studied for their role as entrepreneurs, the driving force of innovation. Beginning with Schumpeter (1939), various studies have been undertaken on the entrepreneurial role of small firms (Rothwell and Zegveld 1982, Audretsch 1995). There is yet to be developed a systematic, theoretical understanding on the relationship between existing SMEs and entrepreneurship. More specifically, there is a fundamental contradiction between the development of hierarchical industrial organisation and the promotion of entrepreneurship, which is still largely unresolved in theoretical terms. For instance, Schumpeter's theory of creative destruction and the role of the entrepreneurs as the primary agent of change is at odds with flexible yet structurally stable industrial networks.

Scholars agree that theories on entrepreneurship have always sat uncomfortably within the realm of economics. Some have argued that SMEs cannot be equated with entrepreneurship, as successful entreprenuership invariably and eventually turns into large corporations (Harrison, 1994b). Some even contests that the concept of inter-firm networks has replaced individual firms as an identifiable scope of research, making the debate on the size of firms effectively obsolete (Teece 1992). As the boundaries of firms become increasing fuzzy through widespread corporate strategic alliances at various stages of R&D, production, marketing and distribution, the debate over firm-size and its relationship to innovative behaviour, technological progress and economic growth has become increasingly complex.

The notion of industrial networks, which imply that the relationships between firms are more horizontal and democratic, has been theoretically attractive and politically promotable by policy makers around the world. However, the reality of industrial networks also involves highly asymmetrical power relationships as discussed in Chapter 9. In Japan asymmetric power relations can be observed between large corporations and sub-contractors, between firms and their employees, and among employees by educational attainment, age (not necessarily experience or skill attainment) and gender.[16]

Can the vast numbers of SMEs in Japan serve as a resource base for the growth of entrepreneurship? Can firms that have long served as sub-contractors in a hierarchical industrial system be turned into entrepreneurs in a competitive market environment? The major challenge for Japan's policy makers is, therefore, to reconcile entrepreneurship with industrial networks, which often involve pervasive hierarchical and inherently unequal relationships between large and small firms.

Although the policy emphasis on entrepreneurship is new to Japan's SMEs, entrepreneurship has a long and well-documented record in Japan's economic history. Today's corporate giants, such as Matsushita, Sony and Honda, emerged in the past as SMEs. During the latest recession in Japan, however, the rate of entrepreneurial expansion was muted. In fact, for the first time in postwar history, Japan's business start-up rate dipped below its closure rate in the mid-1990s. The number of business start-ups in Japan today is approximately 100,000 annually, while the same figure for the US is 740,000.

The history of SME policy in Japan suggests that policies may have long neglected the entrepreneurial side of SMEs while over-emphasising their role as sub-contractors to large corporations within a highly hierarchical system (Aoyama and Teitz 1996). During the high-growth period, this unique relationship between large and small firms was heralded as one of the reasons for the success of the Japanese economy (Friedman 1988, Gerlach 1992, Dicken et al 1997). However, the development of the SME sector as a group of sub-contractors catering to large corporations over the post-war period has arguably eroded at least some of the entrepreneurial spirit which was formerly present in the Japanese economy. Yet, the question remains as to whether reality is that simple since it has been claimed that entrepreneurship can survive and persist even if an economy becomes bureaucratic and hierarchical. The issue then is what has exactly happened to Japanese entrepreneurship? This question warrants further research into SME policy development with a focus on entrepreneurship.

The promotion of entrepreneurship within the existing hierarchical context in Japan has proven to be a trying task, despite a significant change in policy direction in the 1990s. The declaration in 1990 by the Small and Medium Enterprise Agency (SMEA), which states that SMEs are the source of innovation and creativity, is a fundamental shift in policy makers' views of SMEs. SMEs had previously been considered to be the burden, the buffer, or the victim of the Japanese economy, often in need of extra technical, managerial and financial support for their survival. The new view represents SMEs as the primary agents that bring in endogenous renewal and entrepreneurial dynamism to the economy.

One of the first major initiatives by the SMEA to revive entrepreneurship among SMEs was launched in 1995 - an Initiative to Promote and Assist Innovative Small and Medium Enterprises. During the same year, MITI gave permission for employers to offer limited stock-options to their employees, applicable only to start-ups, in order to increase financial incentives for entrepreneurs (SMEA 1998). It will launch a programme modelled after US Small Business Innovation and Research (SBIR) in the 1998 fiscal year.[17] This programme is intended to provide a newly co-ordinated effort by MITI to bring together efforts by various ministries and provide a combined pool of about US $400 million. The SBIR programme was conceived to counteract the lagging industrial competitiveness of the US economy against Japan in the early 1980s, and focused particularly on issues of product innovation. Between 1982 and 1997, the US government diverted a portion of government research spending to SMEs and new start-ups which amounted to up to 2.5 per cent of the total budget (or US $1 billion), funding 5,000 projects in 1997.

Under this programme, investment totalled US $7.5 billion and funded some 46,000 projects. Before the US launched its SBIR programme both US and Japanese governments targeted for R&D funding at SMEs at about the same level - about US$80 million annually. But, by 1997, Japan's funding had fallen to US$50 million, or one-twentieth of the US level.

In conjunction with the SMEA's initiative, numerous regional initiatives have occurred in Japan with increasing local autonomy. Okayama Prefecture in the southwestern part of Japan, for instance, started a programme called 'Young Edison Development Assistance Initiative' in 1995. The programme, named after the famous American inventor, provides assistance to young researchers and graduate students who are interested in commercialising innovative ideas, and has an annual budget of approximately US$1 million (Nihon Keizai Shimbun 1998a). Saga Prefecture as well as Fukuoka Prefecture, both in Kyushu Island, have also launched similar programmes with similar budgetary commitments. These programmes have age limitations (35 years or younger), and all are specifically geared toward researchers in sciences and engineering, generally at local universities. While these programmes reflect new locally based initiatives with new directions, their impact so far has been relatively limited. Okayama Prefecture's initiative, for instance, has funded only 5 projects during the first 3 years of the programme's operation.

Additional initiatives have been developed to strengthen ties between university research and the private sector. Most major universities in Tokyo have set up separate corporations, which provide services for faculty and students, including assistance in the commercialisation of new technologies and the establishment of intellectual property rights. Numerous universities have also launched seminars for current students as well as older adults, to instilling them with a view that entrepreneurship can be a viable option for creating a livelihood.[18] There is even a planned initiative for an entrepreneurship summer camp for elementary school children, to get future workers oriented toward entrepreneurship at an early age. This is quite a dramatic reversal of the previously dominant trends of the post-war era, in which the elites sought employment in the largest and the most stable corporations to secure stability, prestige and social acceptance. Particularly since the 1970s, entrepreneurs have typically been viewed as reckless mavericks, and regarded with suspicion.

To promote entrepreneurship effectively in such social circumstances, some firms have also encouraged entrepreneurship within the firm. These initiatives promote entrepreneurial behaviour among workers within the safety net of lifetime employment (Whittaker 1997). Managers in Japanese companies are also encouraged to urge 'window-side' or sidelined office workers to start spin-off ventures and to exploit new business ideas.[19] The notion of "intrapreneurs" is not a uniquely Japanese notion, however. Such concept has been widely discussed elsewhere as an alternative to independent entrepreneurship to create growth (Rothwell and Zegveld 1982). Recently, this option has become increasingly popular in Japan, and a number of large corporations have established offices within their organisations to promote intrapreneurial behaviour.

These top-down strategies, however, have not been widely successful (Japan Center for Economic Research 1999). Although the media have dubbed 1998 as the third wave of entrepreneurship of the 1990s, a consensus has emerged among Japanese scholars that these waves of entrepreneurship are more accurately described as policy developments and media hyperbole, rather than an actual surge of entrepreneurship.[20] Reports have shown that despite numerous new programmes and policies, entrepreneurs have not come forward in significant numbers. Newly established venture capital companies have been forced to reorientate their lending policies in 1998 as bankruptcies among start-ups have risen.[21] In Japan, even venture capital firms are subsidiaries of large financial institutions, set up specifically to deal with business start-ups. Thus, Japan's venture capital firms are constrained by the current demise of their parent institutions, and are restrained in their risk taking and their aggressiveness in investing in new business ventures.

The largest obstacle for the growth of entrepreneurial activity in Japan is the lack of appropriate labour force. There is a serious shortage of entrepreneurs as well as promoters of entrepreneurs. Potential entrepreneurs lack either the necessary technical skills to provide a new product, or they lack appropriate management and marketing skills to turn a new idea into a viable business venture. The labour market strategies that supported Japan's post-war growth, such as lifetime employment and a seniority pay scale, worked against the development of a dynamic external labour market. The resulting labour immobility has in part reduced the number of workers who consider entrepreneurship as a realistic employment option. The dominance of the keiretsu structure and intra-keiretsu trade has also reduced opportunities for new entrants into business to succeed in the market. Furthermore, a recent survey has pointed out that start-ups in Japan focus on incremental innovation within existing industry and market structures, rather than on emphasising a departure from existing products and services (Japan Center for Economic Research 1999). In other words, few start-ups are modelled after Silicon Valley-style businesses, with aggressive interest in entirely new products. Finally, entrepreneurial individuals are leaving Japan to set up companies in overseas locations, such as Hong Kong, Los Angeles, and Silicon Valley. High operating costs (labour, corporate taxes, rent, etc.) and rigid business practices are the two most frequently cited reasons for establishing a firm outside Japan.

In terms of policy effectiveness, the major shifts in policy direction have not necessarily translated into new approaches. A closer look at programmes and policies reveals that actual policy instruments do not reflect a dramatic departure from those used in previous programmes. For example, to encourage entrepreneurship, business-to-business relationships are promoted to facilitate market development in new businesses. While such programmes supposedly correspond with up-to-date developments in management theories, which emphasise inter-firm networking, they are in fact no different from measures used to promote sub-contracting which were actively encouraged in the post-war period.

Conclusions

Japan's SME policy has had a long history of explicitly sectorally based programmes, the promotion of collaboration among small firms, and the promotion of relationships between large and small firms. Japan's policy has consistently favoured, protected, and promoted the formation of a complex industrial hierarchy. Japan's SME policy has functioned as an instrument that has shaped its unique industrial systems. Active promotion of inter-firm collaboration, along with sectorally targeted policies served as an effective instrument in developing industrial networks that are tight-knit and sufficiently flexible to maintain competitive strength in the international economy.

Such policies have strengths and weaknesses depending on the economic contexts within which they operate. Past events show that policy has played a role in the development of prosperous SMEs within the Japanese economy. Policies can, in turn, influence the shapes, forms and types of industrial networks, given the appropriate economic circumstance that allow policies to be effective. Granted, the same policies that are successful in Japan may not work elsewhere. Such policies work best in an environment where strong government initiatives are accepted and where close business-government relationships exist.

The same policy may not continue to work in the future, even for Japan. The Japanese economy has been undergoing a major transformation through globalisation, suppressed domestic market demand, and the effects of the financial crises of East and Southeast Asian countries. The power and the authority of the government have gradually diminished through social change and the dismantling of political stability since 1993. Globalisation has brought about heightened competition, a massive move of production capacity offshore by large corporations, which has disrupted sub-contracting relationships.

Policy emphasis during much of the post-war period on sub-contractor development rather than entrepreneurship development has led to the stagnation of the renewal potential of Japan's SMEs. SMEs have become dependent on parent firms for financial support, technological diffusion and market development. Without marketing experts and with the limited access to global market information, SMEs are particularly disadvantaged. While many SMEs are struggling to survive, only those firms that have successfully adopted a diversification strategy, through the identification of niche markets, have experienced growth during the past few years. Case studies of these successful firms show that typically they make use of internally accumulated know-how and skills to develop new intermediary industrial products, or devise new technologies to make use of previously wasted by-products. The long-standing craftsman tradition among SMEs still provides an emphasis on manufacturing, and thus existing SMEs have little to do with the new digital economy and cyberspace commercialism dominated by US start-ups.

PART II

THE ROLE OF SMALLER FIRMS IN COMPLEX PRODUCTION SYSTEMS

4 Challenges and Pathways for Small Sub-Contractors in an Era of Global Supply Chain Restructuring[22]

Poul Rind Christensen

Introduction

Serving industrial markets as a sub-contractor seems to be an important position taken by small and medium sized enterprises. Data for Europe (European Commission 1994) indicate that between one third to more than half of the SME population act as sub-contractors. A Danish in-depth study indicates that approximately 45 per cent of small and medium sized manufacturers are sub-contractors[23] (Andersen and Christensen 1998).

The market for sub-contracted goods is becoming increasingly international, partly caused by improved communication infrastructure. It is also triggered by a growing number of global production networks controlled across national boundaries by transnational corporations. At the same time it is triggered by a specialisation of activities in production. Large international firms sub-contract - or out-source - larger shares of their value added than before. This is mirrored in patterns of international trade in semi-manufactured goods (Lüthje 1999, OECD 1992).

There seems to be little disagreement about the role taken by transnational corporations in the organisation of what is often labelled global[24] supply chains. Still, there seems to be disagreement on how these changes influence small and medium sized sub-contractors and the local industrial communities in which they are embedded. In some studies these sub-contractors are seen as the cornerstones of a future flexible organisation of global supply chains (Sabel 1989, Scott 1988a, Perrow

1992). In other studies (Amin 1989, Scully and Fawsett 1994, Rainnie 1993) small and medium sized sub-contractors are seen as being much more squeezed.

The purpose of this chapter is to highlight the changing opportunity structures imposed on small and medium sized sub-contractors by the restructuring of global supply chains. Propositions will be developed on the challenges and pathways faced by small and medium sized sub-contractors, when the supply chains in which they are positioned become internationally orientated and even evolve into a system co-ordinated by a few global players.

The on-going transformation of the global structure of production also implies a break-up of mainstream analytical perspectives and frameworks of interpretation. View on small and medium sized sub-contractors' positions in supply chains differ according to the analytical perspective taken. They may be classified on their positions in global supply chains or on their position in regional and national production systems. In this chapter different types of sub-contractors are identified and discussed. It is argued that changes in competitive conditions will alter mechanisms for selecting sub-contractors in an uneven way, with the likely consequence that different types of sub-contractor can expect to face different opportunity structures.

Based on propositions linking the typology of sub-contractors to differences in their spatial position, implications for different types of SME sub-contractors are discussed. Some of the spatial implications for small and medium sized sub-contractors will also be considered. To highlight some of the patterns and propositions that are proposed, results from a recent empirical study of manufacturing sub-contractors in Denmark are used to illustrate key issues.[25]

Changes in the international division of labour - evolutionary perspectives

The international orientation of OECD economies has gone through three major phases, each of which has impacted differently on the character of the international division of labour and global competition.

The first phase occurred in the early post-war years. It was characterised by a substantial growth in international trade mainly in raw materials and final products. This phase can be labelled 'the simple export phase'. The expansion of international trade was triggered by the reduction in tariffs and other import restrictions and the creation of multilateral trade agreements. This phase, sometimes called the golden age of trade, lasted until the end of the 1960s (OECD 1992).

International trade in this period was founded on old patterns of factors endowments and lead to a widening of the horizontal division of labour, i.e. a stronger mutual specialisation among nations inside specific lines of industry and specific product groups.[26] These patterns of specialisation have been analysed in a number of studies. Dalum and Villumsen (1996) characterised the specialisation pattern as 'sticky' and found that it had been stable among OECD countries since the 1960s.

This sticky pattern of specialisation also to influenced the second phase of internationalisation. This was the phase of Foreign Direct Investment (FDI). Although it occurred in the pre-war period, FDI grew considerably in the 1970s. One of the implications of this growth was that the share of international trade controlled by multinational enterprises grew to hitherto unknown levels. At the end of the 1980s, their share was more than 60 per cent of world trade (OECD 1995), indicating the growing importance of a small number of global competitors, global branding and the control of global supply networks. While the first phase of internationalisation can be explained in terms of comparative advantage, this second phase was concerned more with the exploitation of ownership advantages[27] through the control of specialised assets. A few multinationals dominated competition in a number of product markets. Among the most substantial forces behind the FDI of these multinationals was the drive to gain market access and the advantages to be gained from combining production factors from several continents. These developments produced the seeds of the third phase of internationalisation - the phase of globalisation.

This third, globalisation phase has been marked by a shift in the international division of labour. In this phase, the multinationals are said to have changed their organisational practices, and thus the orientation of foreign direct investment (FDI), trade and competition. While previously FDI hitherto had been organised on a market to market (country to country) base, much stronger functional integration across national boundaries has now developed in some corporations, creating 'transnational corporations' (TNCs). Diversified forms of co-ordination as well as centrally co-ordinated systems are seen operating side by side. Although these corporate networks of divisions, production units, R&D units and sales units are organised in quite different ways, they also share some common characteristics (Hedlund and Rolander 1990).

Their configuration of international activities is based on new views on organisational boundaries which favour outsourcing, activity sharing and the control of markets through strategic alliances and network based sourcing, distribution and sales. A common theme in the modelling of TNCs is that they try to learn and gain from opportunities in their global environment through the organisation and management of their activities. (see Dicken 1998, Sölvell and Zander 1991, Harrison 1994b and Tödling 1994, for diverse yet comprehensive interpretations of the globalisation of the international production system).

The transition from the first phase of international trade to the third phase of globalisation has been marked by a remarkable, but often overlooked, shift in the international division of labour. While the export phase was characterised by a high level of horizontal specialisation, the phase of globalisation is marked by strong growth in the vertical division of labour. In this process of vertical specialisation, the meaning of 'country of origin' for specific products is becoming increasingly meaningless. Increasingly, products are the result of production and other types of value adding activity from many parts of the world, and the same product and label is often assembled simultaneously in several parts of the world. This change brings to mind Hirschmann's (1958) work on economic development in which he found a strong relationship between the growing complexity in the linkage structure in an industry and periods of transformation and growth in the economic system as a whole.

The change from a horizontal to a vertical perspective on the division of labour has been accompanied by strong growth in the international trade of semi-manufactured goods and components (OECD 1992, 1996a). Indeed, Feenstra (1998) has highlighted the close relationship between the growing integration of world markets and the progressive disintegration of production processes. While Feenstra has discussed technical disintegration, i.e. the ability technically to separate tasks, disintegration also has managerial as well as geographical dimensions.

Major changes in sourcing patterns

Sub-contracting has traditionally been viewed as a local, regional or national phenomenon. It is a view reflected in the spatial interpretations and subsequent discussion of Perroux's growth pole theory from the 1960s.

The internationalisation of markets for semi-manufactured goods and sub-contracting now constitutes a major transformation pressure affecting small and medium sized sub-contractors who are used to serving local and national markets. The pressure is the same even when they remain domestically orientated and do internationalise alongside their customers. This is so because sub-contractors still have to conform to the international standards of technology, quality, delivery and after sales service that evolve in international networks.

This issue is also of relevance for the study of clusters of small sub-contractors, since the routines and practices developed locally - or in national production systems - evolve into the collaborative traditions that constitute organisational communities (Aldrich 1999). These traditions then put under pressure by the administrative managerial practices set by transnational corporations that operate in those same places.

However, since local features have dominated in most instances of sub-contracting, the institutional settings developed in supply chains also differ strongly from country to country. Even inside the same trade, huge differences can be found in the co-ordination and control of activity chains. The configuration of supply chains and the collaborative practices developed in Japan have thus differed strongly from those developed in western societies, as discussed in the previous chapter. The US tradition of vertically integrated production systems, as discussed by Chandler (1977), may in part be explained by the political and legal practices regulating firm collaboration (Jorde and Teece 1989) and, consequently, sub-contracting practices in the US.

Strong selection mechanisms have worked to bring major change in production best practice in a number of industrial sectors. The Japanese sub-contracting system in particular has served to demonstrate that Western business systems suffered from severe organisational failures. Perrow (1992) has emphasised that this is caused by the inability of top management to cope with small-numbers bargain positions. Management of corporate organisations has to deal with the problem of bounded rationality arising from internal vested interests, asymmetric information and impenetrable information. Different organisational solutions to cope with inflexibilities and deficiencies in co-ordination in corporate organisations have been proposed in the contexts of U-form (Unified) and M-form (Multified or divisionalised) organisations. The out-sourcing of tasks to external suppliers can be interpreted as a managerial response to the malfunctioning of internal co-ordination and governance. Out-sourcing is thus a fundamental choice mechanism in growing corporate organisations. But, the momentum changes when production facilities and activities become globally dispersed.

In many sections of industry a second wave of supply chain development appears to be underway. Now, global supply networks are gaining in importance at the expense of traditional, nationally based supply systems. In a large volume of literature, change in best practice has been interpreted as a shift away from vertically integrated production to a system dominated by out-sourcing, and from arms-length relations (exit) to lean relations (voice) (Helper 1993, Morgan 1997, Hines 1994). These changes influence the positions held by firms, especially by small and medium sized sub-contractors, in supply chains. The present state of the Japanese sub-contracting system illustrates the problems faced by SMEs when contractors globalise their network of production facilities.

Convergence or divergence in the configuration of supply chains?

As suggested in Chapter three, there appears to be a consensus that superior Japanese productivity has at least five underlying causes:

- lasting relationships facilitating long-range planning and investment in those relationships, as well as a strong mutual understanding of roles in the supply chain;
- the building of clusters of sub-contractors in close proximity to contractors (Toyota City and the evolving Technopolis programmes in Japan are often quoted examples);
- a labour market that tolerates lasting differences in wage levels between large and small enterprises;
- strong public regulation and support of relationships between contractors and sub-contractors that reduce opportunistic behaviour and sustaining investment and improved practices; and
- the inclusion of sub-contractors in innovative activities.

However, it is a paradox that while the Japanese system of sub-contracting and its key practices have served as a model for industrialists all over the globe, there are strong signs of erosion in the institutional and territorial features of the system in Japan. The Japan Small Business Research Institute (1995) thus reports a growing crisis among small and medium sized sub-contractors in Japan: a crisis with its origins in more than the current economic problems being experienced in Japan. The crisis seems to have basic structural features related to the ongoing globalisation of the Japanese production system. It is thus paradoxical that while European and US corporations try to learn from Japanese supply chain management practices, the Japanese supply system is itself undergoing major modification to incorporate exit characteristics similar to those seen in western supply systems (i.e. more open and price oriented and less dependent on local suppliers base in the home market).[28]

With Rainnie 1993, it is tempting to conclude that formerly huge and territorially based differences in practices of supply chain management are tending to level out. However, other contributors emphasise that the way transnational corporations configure their supply activities is in no way uniform (Carter and Narasimhan 1996). They may be organised along lines defined by the divisional (M-form) structure or in other cases in a more centralised U-form where global co-ordination may take place.

From a dynamic perspective evolving patterns of sub-contracting are the result of the interplay between, on the one hand, a variety of different corporate based supply chain management (SCM) practices and on the other hand, different traditions institutionalised in a territorially based organisational community. It is in this environment that different managerial traditions and governance mechanisms meet. This can be exemplified by a formalised, rule based corporate supply management system meeting an informal, socially based network of personal relations, revolving around everyday transactions in a local industrial frame.

Bearing in mind such complex dialectics, it is helpful to clarify the 'changing rules of the game' by highlighting a key aspects of those changes in sourcing practices which seem to develop into generic traits in the globalisation of corporate based supply chain management:

- *strategic sourcing* gains momentum;
- *rationalisation of the supplier base* is widespread; and
- *segmentation of supplier relationships* is an evolving practise;

These key aspects of supply chain management are used to structure the discussion in the following sections of the chapter. Against the background of the changes in the use of sub-contractors discussed under these headings, a clear need emerges to construct taxonomy of sub-contractors to match the evolving collaborative regimes within which they are involved. This taxonomy is developed in a separate section of the chapter. In a final section, 'the spatial opportunity structures' available to different types of sub-contractor are discussed.

Strategic sourcing

It is generally recognised that sourcing amounts to more than mere purchase. Selecting suppliers that can make strategic contributions to business development has become a key issue. Among the reasons for this upsurge in interest in sourcing is intensified knowledge competition, which forces enterprises to reduce lead times in product development and to master more and distinct areas of knowledge. At the same time, the international operations of corporations are resource demanding. Supply is no longer a matter solely of current prices and the costs of procurement. Increasingly it is seen as a supporting the specialised technology and knowledge base of a contractor.

In an assessment of the procurement practices of 149 US corporations, Birou and Fawsett (1993) found that the intensified competition over technology was central to the growing use of specialist, knowledge-based sub-contractors. They found that in almost 50 per cent of enterprises the motive for international sourcing was based on the fact that the desired knowledge was obtainable only abroad. Only 25 per cent sourced abroad because of better quality or more advanced technology. A cautious interpretation points to the conclusion that access to specialised, tacit knowledge may be decisive in promoting the future growth in international procurement.

Therefore, supply chain management tends to involve more in-house departments and functions to select suppliers and to administer procurement procedures. When supplies have a critical technical value, which is not replicable, or the suppliers are actively involved in the specifications of the supply, then a contractor is dependent on their supplies and may be eager to safeguard supplies and support stable relationships. In other words, these sub-contractors give strategic value to the contractor. This is partly due to rising exit costs and partly because value can only be extracted in the long run.

A study of Danish sub-contractors revealed that more than 70 per cent of the relationships they had with their four most important Danish customers were more than 5 years old. When it came to their foreign customers, more than 60 per cent of the relationships were more than 5 years old.

Many specialised suppliers support knowledge and competence building with the contractor they serve. If contractors want to extract value from their own core competencies, they also rely on this 'supplied competence'. Specialised supplies are often developed through close co-operation between contractors and sub-contractors. In this way, a knowledge base crossing organisational borders is built, which will fall apart if the relationship is broken (Christensen 1988). Many suppliers thus gain strategic value in the sense that relationship building has demanded significant investment.

Rationalisation of the supplier base

Currently, there is a strong tendency among firms to externalise activities and to use external suppliers (Kearney 1993). It is, therefore, of strategic importance to the contracting enterprises that the overall base of suppliers is co-ordinated, and operates in a cost efficient way.

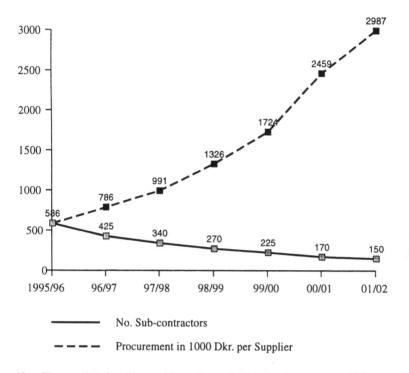

_____ No. Sub-contractors

- - - - Procurement in 1000 Dkr. per Supplier

Note: The example is from Glunz and Jensen International, production manager H. Ejsing (1997)

Source: Andersen and Christensen, 1998

**Figure 4.1   Numbers of sub-contractors and levels of procurement.
Development and prognoses**

Therefore, a prime consideration in supply chain management is associated not with the efficiency of bilateral relationships but rather with the overall efficiency of the whole supplier base. For a contractor a basic strategy could be simply to reduce the number of suppliers dealt with directly, thus reducing the number and variety of relationships to be managed. This strategy is demonstrated in Figure 4.1, illustrating the trends and policies of a Danish contractor, Glunz and Jensen International. This firm reduced the number of sub-contractors it uses by almost half between 1995 and 1998. At the same time, procurements from each supplier have, on the average, more than doubled. Underlying this trend, a rationalisation of sub-contractors has also taken place.

The trends suggested by this case study are supported by Kearney (1993) in a study of the sourcing policies of 1000 European transnational corporations. The study pinpointed the major cost savings that can be achieved when supplier bases are rationalised. The savings include:

- benefits from economies of scale obtained by the suppliers;
- reduction in the number of relationships to be handled;
- benefits from closer relationship with fewer suppliers; and
- economies of scope in the co-ordination of supplies.

Reductions in the supplier base tend to put pressure on small sub-contractors. They are forced either to expand their production capacity or to launch a collaborative venture with other competing sub-contractors in a joint effort to scale up supplies. The effects of supplier reduction programmes are reinforced in the cases where contractors have a policy of not being responsible for more than a minor share of the production capacity of a sub-contractor. In such cases, minor sub-contractors are often forced to expand their portfolios of customers if they want to do 'business as usual'. Small suppliers may also be squeezed by the demand for them to have a logistics capacity so they can serve global procurement networks. Equally, they may be squeezed by new demands on their managerial capabilities, and the requirement to contribute to the development of new generations of products.

Reductions in their supplier bases are primarily achieved by contractors in two ways: replacement of multi-sourcing with single sourcing, and through the use of systems suppliers. The demand for system suppliers is related to their ability to co-ordinate a number of related component producers to develop, produce, assemble and deliver a whole sub-system instead of single components. What is in demand in addition to physical supply is a managerial capacity to select and co-ordinate other suppliers, i.e. the ability to co-ordinate their own tasks with those of related component producers.

Through the rationalisation of the vertical supply system, contractors force sub-contractors to 'in-source' new, related production tasks or to initiate and manage horizontal collaboration to achieve the supply of whole sub-systems. This capability may involve the development of both joint design and a closer technological interface.

Vertical activity rationalisation thus implies a stronger and tighter horizontal task partitioning, which in turn rests on the ability to build collaborative regimes in favour of long-run investments across organisational borders. This pressure for horizontal task co-ordination tends to favour related sub-contractors, located in close proximity, with a developed organisational community of supporting institutions, and traditions favouring collaborative ventures.

A final point of importance in the rationalisation of the supplier base is the process of sub-contractor selection. On the one hand, the malfunctioning of sub-contractors is costly. On the other hand, it is costly to change to new sub-contractors. Time and investment is needed to establish routines with new suppliers, and there are search and auditing costs involved in finding new sub-contractors, not least when the search has to be undertaken internationally. Therefore, many transnational contractors have developed sophisticated step-wise and differentiated auditing procedures.[29] These step-wise auditing procedures are demanding of small sub-contractors' abilities to document their technical and managerial capabilities. In very stable supply systems, competition tends to be displaced at the stage of auditing (pre-evaluation), where it becomes difficult to break into established supplier relationships. In these situations, traditional market competition based on 'exit' is outdated. Instead, contractors establish competitive regimes of an administrative character. Such administrated competitive regimes are framed by auditing procedures, rules of yearly cost reductions, and the imposition of benchmarking procedures on established suppliers.

Segmentation of supplier relationships

It is important, therefore, for contractors to build different competitive regimes with different types of sub-contractors so they can secure the key competencies wanted from each of them. At the same time, collaborative efficiency must be sustained through relationship building with different types of sub-contractors selected.

A move away from a strong dyadic focus may well be justified by the simple fact that contractors handle not one or few suppliers, but most often use huge numbers of suppliers with strongly diversified aims and, consequently, huge differences in interaction. Diversified relationship building can be viewed in portfolio terms, with different sets of sub-contractors occupying different strategic positions in their relationships with contracting firms. Therefore, instead of seeing relationship building as a one-dimensional phenomenon evolving from 'exit' to 'voice', contractors can be viewed as building a mix of linkages with sub-contractors based on the types of relationship they seek.

A fundamental feature is that contractors tend to segment their supplier base into different interactive regimes. Each regime is characterised by its own pattern of collaborative practice and administered competition. From this perspective, it is evident that the creation of the portfolio of relationships has a strategic importance of its own. Through the segmented portfolio, the contracting enterprise can establish diversified managerial guidelines for auditing, for resources devoted to specific types of relationships, and for relationship building with different types of suppliers.

Also, at the level of the portfolio, differences among TNCs as contractors show up. Some TNCs emphasise central co-ordination in auditing and procurement, while others make these matters a divisional responsibility. The issue of procurement also differs fundamentally between single location contractors and contractors with multi-local production units.

Although supplier segmentation may vary greatly, it has two essential and important characteristics. First, some suppliers may be crucial to a contracting enterprise because of their special, day-to-day operational capabilities or because of their capabilities in terms of innovation and new product generation. Second, some suppliers play a key role because of the sheer scale of their operations.

Based on these two dimensions the concept of relational regimes can be developed. In Figure 4.2 four relational regimes are mapped. Each is characterised by different buying centre arrangements;[30] specific auditing procedures and different guidelines on interactive practice. Guidelines on internal co-ordination in the contractor's procurement network may differ as well as guidelines on resource commitment.

While relationships in the two upper quadrants of Figure 4.2 are basically dialogue orientated and require a strong mutuality in relationship building, interactions in the lower quadrants are more likely to be rule based. The mapping approach can also be used to show that different types of learning circle are established between contractors and sub-contractors.

This schematic portfolio model also shows that small sub-contractors may have very different relationships with their contractors. It follows from the logic of Figure 4.2 that small sub-contractors tend to be positioned in the two left quadrants of the diagram. Although small sub-contractors often act as, and are treated as, critical suppliers, it is probably only in a minority of cases that they participate in formal R&D activities. Most often, the use of sub-contractors is based on their innovative and technical capabilities, which by its nature is an entrepreneurial capability. Even simple operational suppliers may gain critical value through a flexible delivery service or a logistics capacity. On the volume side, it is quite evident that small sub-contractors are limited in their capacity to satisfy the high volumes required by large mass producing contractors, although high volume supply is a flexible concept.

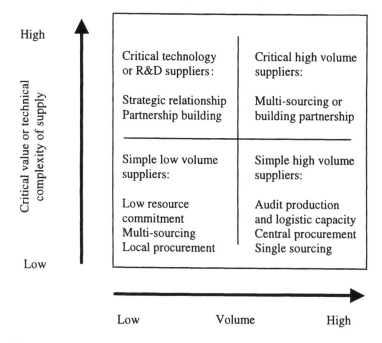

Figure 4.2   Segmentation of relational regimes

A new taxonomy of sub-contractors

The evolving complexity in international supply chains, and ongoing change in supply chain management, has imposed a diversity of new demands on sub-contractors. There is a need, therefore, to elaborate and refine the concepts used to understand this diversity of new roles.

In supply systems, in which tasks of non-strategic importance are the only ones externalised, a sub-contractor has been defined as an independent enterprise carrying out operations according to specifications and plans given by the contractor (Holmes 1986). The custom-made character of the relationship is at the heart of this conceptualisation.

In order to match the evolving variety of roles placed on sub-contractors and the variety of relationships evolving in international supply chains, a differentiated typology of sub-contractors has evolved (Blenker and Christensen 1995). Basically, the typology is made up of five types of sub-contractors. For the sake of simplicity, this typology can be reduced to three categories. This consolidated typology is shown in Figure 4.3.

The fact that a specific activity is sub-contracted does not necessarily imply significant co-operation and co-ordination between the two parties involved. However, some tasks do necessitate significant co-operation and co-ordination in order to be fulfilled. Thus, co-ordination is a basic dimension of the taxonomy.

Also underpinning the taxonomy of sub-contractor interactions and relationships is the issue of task complexity. Some tasks are so simple or so standardised that they are relatively easy to co-ordinate across organisational boundaries. Other tasks are more complex. Demonstration may be needed, or case-by-case adaption and adjustment might be required. The interface between contractor and sub-contractor may be difficult to co-ordinate. Co-ordination might involve time, considerable effort and investment to establish appropriate routines. Far greater resources are devoted to this type of relationship.

The three groups identified in Figure 4.3 help us to visualise the positioning of traditional sub-contractors between, on the one hand, sub-contractors engaged in the development of new products and processes with a contractor and, on the other hand, those specialising in large scale production of standard, yet specialised, components, manufactured at low cost.

The 'traditional sub-contractors' comprise two sub-groups. The first sub-group, the 'capacity suppliers' are characterised by their role as buffers and cost efficient suppliers. Most often they act as flexible supply units safeguarding the optimal capacity utilisation of their customers. It is these sub-contractors that are seen as reservoirs of large firms (Semlinger 1993). They are often located in close proximity to their contractors. Since their customers have the capacity and the skills to handle the tasks offered, their most important competitor is actually their customer. This type of supplier is exposed to the make/buy decisions of its customers. Exit costs are low.

The second sub-group of 'traditional sub-contractors' are the 'specialised suppliers'. They are characterised by their co-specialised competence vis-à-vis the customers. Their competitive situation differs from that of the 'capacity suppliers'. They operate through instructions and prescriptions set up by their customers (see for example Dicken 1998). Sometimes the customers even deliver the materials on which the sub-contractors perform pre-specified operations (functional sub-supplies). In such cases, close proximity is needed for simple logistical reasons.

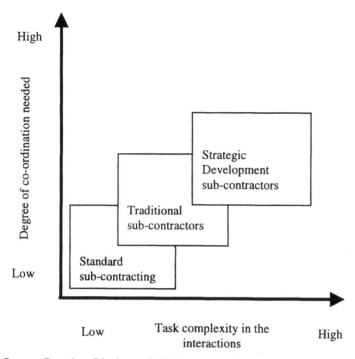

Source: Based on Blenker and Christensen, 1995

Figure 4.3 A taxonomy of sub-contractors

Information exchange and co-ordination for 'traditional sub-. contractors'. Operative complexity is relatively low and the level of contractor specification is high. At the task-complex end of the 'traditional sub-contractors', mutual specialisation between the parties expands, although relationships remain simple in structure. Exit costs tend to rise owing to the level of investment and mutual adaptation in the relationships. Highly stabile relationships are created through the exercise of 'management by exception'.

Some sub-contractors have invested heavily in the specialised production of standard components, for incorporation into a final product. As a type, such 'standard sub-contracting' is favoured because of the economies of scale it affords in international markets (Figure 4.3). The economies can be obtained through high levels of product and task standardisation, together with low interactive complexity, and low adaptation to individual customer needs. Basically, standard sub-contractors are less vulnerable to rationalisation in the supplier base than are other types of sub-contractor. Therefore, standard sub-contracting enterprises are placed at the lower, least task complex, end of the typology.

In the box 'strategic development sub-contractors' in Figure 4.3, two types of sub-contractors are found. One sub-group represents sub-contractors of strategic though not exclusive value to contractors. Their value may reflect their specialised technological capability, their ability to provide specialised after sales service, their access to co-specialised suppliers, and their ability to act as co-ordinating system suppliers, which may also be of strategic value. Also, their ability to contribute to different stages in the product development of a contractor is of value. Therefore, in relation to this type of sub-contractor, exit costs are high, but not necessarily in a balanced way. The second sub-group of 'strategic development sub-contractors' have partnership based relationships characterised by even stronger mutual strategic value and high exit costs. These relationships are defined by strong mutual dependency caused by the complexity involved and the levels of investment that have been made. They are also characterised by an element of strong mutual destiny because the capabilities and business opportunities of both contractors and sub-contractors are closely interwoven. The sub-contractor is thus often highly involved in joint R&D activities or product development with the contractor.

As indicated in Figure 4.3, there is a certain degree of overlap between the different types of sub-contracting relationships identified, indicating their mutually competitive character and areas of blurred co-operation. It also shows that a specific level of task complexity may apply to different levels of co-ordination, implying an evolutionary trajectory in levels of organisational complexity. Different routines and practices of co-ordination may develop, and these may match different phases of learning, differences in proximity or differences in the size and diversity of the organisation of contractors.

Changes in small sub-contractors opportunities

Small sub-contractors all around the world are greatly affected by the ongoing restructuring of supply chains. Small sub-contractors' opportunities in the evolving global production networks are only infrequently analysed. Since sub-contractors' opportunities are closely linked to wider industrial dynamics, in this section those opportunities are examined in the light of the theory of industrial dynamics (Carlsson 1987, Lundvall 1998).

From this theoretical perspective change arises from innovation and adaptation triggered by external transformation pressures and structural tensions among elements and sub-systems in the system as a whole. The opportunity structures of small sub-contractors can then be seen as the result of processes of selection and transformation in which technical, organisational and geographical features play a role. Therefore, change in the organisation of supply chains involves a techno-economic dimension, an organisational dimension and a spatial dimension.

The techno-economic dimension of supply chain change provides a basic conceptualisation of the innovative or adoptive changes that take place within them. These changes combine to create specific structural tensions.

In the organisational dimension, the focus is on evolving managerial practices following the reorganisation of supply chains. While earlier sections of this chapter have explored the diffusion of new managerial practices and touched upon questions of convergence and divergence, the organisational issue examined in this section focuses on the changing opportunities of small sub-contractors.

The spatial location and interconnection of activities and flows are often seen as outcomes either of techno-economic tensions or organisational innovations. In the argument developed here the spatial configuration of activities is seen as an input that changes sub-contractors' opportunities.

The techno-economic dimension

The impact of global change on the international division of labour is reflected in the changing linkages of firms within supply chains. Increasingly, these chains are also linked across national and continental territories, forming global production networks. As a result, the diversity of sourcing options has been expanded, leaving those actors configuring specific supply chains with greater freedom in combining suppliers. This variety in sourcing options feeds both productivity and innovative potential. The knowledge base is expanded and becomes more varied.

When the divisions of labour changes, the distribution of knowledge is also changed. Skill learning is made more difficult and risky for small sub-contractors since new competencies are generated and combined in larger networks of actors than they are used to handling. Small sub-contractors, therefore, experience greater ambiguity in their positioning in supply chains, and the roles of sub-contractors become more diversified. Change is also fuelled by the standardisation of components. At the same time, there is also a trend towards diversification of the knowledge base and the customising of supplies.

While large globally operating contractors can improve their productivity through the combination of diverse resources and knowledge bases, many small sub-contractors have to reconsider their positions in the typology that has been outlined in Figure 4.3. While a dominant segment of Danish sub-contractors[31] still operates in the traditional mode, it is among the 'standard sub-contractors' and 'strategic development sub-contractors' that expansion has been greatest. Therefore, a large number of Danish sub-contractors are now under pressure.

The organisational dimension

The pressurised position of small sub-contractors is not created solely by techno-economic forces. Pressure is also caused by changes in the organisational architecture and the concentration of ownership in many sections of industry. Also, managerial practices (including the evolving dominant 'Supply Chain Management Ideal') influence the performance of small and medium sized sub-contractors. They have to learn new managerial practices and learn to live with a 'rationality' set by corporate administrations. They also have to see themselves as part of a larger supply system guided by 'systemic performance measures', and have to learn to take advantage of the corporate support that these provide.

Through their administration of the supply base, large contractors tend to internalise the way competitive forces work, i.e. they develop new ways of administrating competition in the supply chain. Many authors stress that in addition to the old issue of prices, large international contractors now expect their sub-contractors to contribute to the development of the joint efficiency of the supply system. This may take the form of contributing to new product generation as well as the joint administration of daily operations.

It follows implicitly that small sub-contractors face severe transformation pressures. There is an ongoing tendency to push small sub-contractors to the lower tiers in the supply chain. They then have to rely on information from suppliers working between themselves and the contractor - contract producers or lower levels in the supply chain hierarchy that take over strategic authority (Powell and Smith-Doerr 1994).

Exit costs tend to grow with technical and organisational integration. Small sub-contractors, who have learned the lesson not to be too dependent on one or a few contractors, now face a new era of supply chain management in which they are forced to entrust high proportions of their production capacity to a few contractors. It is difficult for small sub-contractors to judge contractors' new signals of 'lean' (voice) management, when they have been used to 'arms length' (exit) strategies. Therefore, many small sub-contractors may well find further investment along this 'lean line' of administration to be risky.

Since small sub-contractors are used to acting entrepreneurially in response to new business opportunities, it can be suggested that long-range administration aimed at building business opportunities is a long way away from their usual way of interpreting opportunities. In this way, localised perspectives on opportunities meet with global, corporate opportunity structures. In such highly administrated supply chains, small suppliers dominated by an entrepreneurial management style, are forced to adapt to routines and interactive behaviour which is much more administrative in character. Apart from possible alienation, the subordination this involves tends to destroy entrepreneurship. We do know, however, that interaction in networks and supply systems may have unintended consequences as far as the contractor organising the network is concerned. So, although we do find such general implications for small sub-contractors it is important to bear in mind the complex dynamics of sub-contracting.

The spatial dimension

While transformation pressures may seem to be overwhelming for small sub-contractors from an organisational and firm-based perspective, opportunities may seem different from a spatial perspective.

Since out-sourcing leads to the relocation of activities and the re-configuration of activity chains, knowledge distribution will change spatially as well as organisationally. There is an urgent need, therefore, to study industrial linkages to clarify the way in which spatial knowledge distribution depends on and co-evolves with task location and the organisation of activity linkages in globalised activity chains.

When out-sourcing grows and suppliers come to play a key role in production, efficiency is still largely determined by the way different production units are interrelated and interact in the system (Gadde and Hakansson 1992). Small sub-contractors' business opportunities in terms of territorial expansion are thus based on the evolving mutuality of business opportunities. This mutuality consists of a vertical element in relation to contractors and a horizontal element in relation to complementary sub-contractors.

The different ways in which global contractors bring about internal co-ordination is also of importance because it influences the patterns and routines developed in their co-ordination of adjunct suppliers. In a divisionalised structure, daughter units may view themselves as competitors of other corporate units. Their position in the corporate system may thus depend on their ability to organise the community of sub-contractors operating in their hinterland.

There has been a long and strongly argued debate on the role of localised, socially regulated networks of small businesses versus the role of the networks configured by large TNCs. An overview of the debate can be found in Pyke, Beccatini and Sengenberger (1992) and in Powell and Smith-Doerr (1994). It is interesting to note that this debate has only briefly considered evolutionary dynamics. Very few analyses exist on how TNCs tap into clusters of small sub-contractors located in industrial districts or similar small business clusters. There have been no studies on the dialectic between the two types of network involved and their different patterns of regulation and co-ordination.

Evolving systems of sub-contracting, as well as the evolving governance structures that dominate those systems, are important analytical links between studies of global production networks on the one hand, and studies of local and regional production networks, on the other. The need for this analytical link is articulated and supported in several contributions. While Morgan (1997) finds a need to research the role of the embedded branch plant, Amin and Cohendet (1999, p.102) find that local subsidiaries sometimes "generate considerable scope for regional asset building". However, their comparative view on the dynamics of local and global production systems is underlined by their argument that it is difficult to claim the superior economic competitiveness of local knowledge environments when the dynamics of local decentralised business systems are compared with those organised on a global scale (Amin and Cohendet 1999, p. 101).

The position taken in this chapter is that, although the entrepreneurial behaviour of small sub-contractors may be over-ruled by strong administratively orientated management practices embedded in corporate systems, these corporate systems are nourished by the same regional supply systems.

This balance of forces influences the opportunity structures developed by sub-contractors. Localised networks may take the form of industrial districts or clusters in national production systems. These networks are characterised by commensalism, symbiosis and patterns of power, which are institutionalised through long exchange traditions, values and practices that evolve within specific trades as well as within public and semi-public institutions. The exchange networks of industrial districts are often characterised by their social embeddedness and their unique and socially embedded interactions (Scott and Storper 1989) i.e. informal socially embedded sanction systems, the informal regulation of competition, and collaboration.

Such organisational communities[32] (Aldrich 1999) often differ in their main features. Some are organised around a specific technological design, others are organised on the basis of a joint pool of qualified labour, while yet others may be organised around a knowledge providing institution. In a comparative study on the contrasting symbiotic relationships in Silicon Valley and the Route 128 region, Saxenian (1994) found that learning cycles and resource generation was much more inward orientated in the second than in the first area.

Although these organisational communities may take very different forms in terms of collaboration, knowledge distribution, the building of supporting institutions, and managerial practices, they do share a number of features in common. First, they all develop an internal regulatory regime, which tends to be unique. Second, the organisations involved share a common destiny in the sense that they are dependent on each other and on the prosperity of the others, although not always on a one to one basis. Third, firms in organisational communities learn in ways that other firms do not (Aldrich 1999, Asheim 1998, Malmberg and Maskell 1997).

Therefore, when specialisation and the vertical division of labour are deepened internationally, it is also paralleled by a deepening of specialisation in regional and national territories. Some sub-contractors tend to evolve into system-suppliers bridging the managerial gap between tradition embedded in local clusters and corporate practices. In other cases, TNCs take over local units. When TNCs specialise their activities worldwide, it has repercussions for the specialisation of SMEs located around the world. The more firms specialise the more they depend on each other. This proposition seems to hold true in a supply chain context as well as in an industrial district context.

This issue of dependency points to a systemic dimension of the internationalization process which goes beyond traditional firm-based views in that competitiveness and thus international orientation are seen very much as interdependent phenomena. Export projects or the development of sub-systems are therefore most often based on related skills and tasks floating in a localised network of sub-contractors. Thus, sub-contractors embedded in regional supply systems depend on TNCs working as 'impannatories'[33] to provide market access, while the TNC depend on the entrepreneurially based flexibility and creativity embedded with actors in the local supply network.

Sub-contractors located in localised business clusters and industrial districts may gain considerable advantages from this vertical pressure. The advantages of joint location for small and medium sized businesses seems so strong that manufacturing industry tends to be increasingly clustered (Malmberg and Maskell 1997).

In the Danish study drawn on in this chapter, it has been shown that the 'spatial' roles of manufacturing sub-contractors differed strongly. More than half of all sub-contractors thus had their main activity founded on relationships both with domestic suppliers and domestic customers. Almost 20 per cent used Danish suppliers but were highly export oriented, while approximately 15 per cent relied on foreign suppliers but domestic contractors as customers.

While these three types of sub-contractor were firmly embedded in a national production network - including international daughter companies located in Denmark - the study also pointed to a group of sub-contractors (12 per cent) with weak links to the national production network. They were, nevertheless, embedded in the Danish labour market, Danish infrastructure, and Danish national institutions.

Given a strong trend towards standardisation in supply chains, there is reason to believe that the future will witness growth in the numbers of sub-contractors in 'local enclaves' in the national production system. They may be seen as signs of disintegration following the expansion of the global production system – global processes breaking down industrial districts and eroding regional production networks.

Perspectives and discussion

Against the background of the supply chain dynamics outlined in this chapter, two activity-based perspectives on small sub-contractors' challenges and pathways are outlined in this section. One is a firm based perspective, the other is a network perspective.

The firm based perspective

From the firm based perspective, small sub-contractor's positions in supply chains are of major interest. Since knowledge and competence tends to follow those who undertake specific activities, the small firm is vulnerable to the mobility of activities within supply chains. Therefore, a major issue for the small sub-contractor is how to make itself indispensable to the efficient flow of activity.

Small sub-contractors, especially the traditional ones with product activities that are easy to copy, may make themselves indispensable through proximity, flexible delivery, and specialisation or by being more efficient. From this perspective, opportunities and challenges depend very much on the type of sub-contractor and the position it holds in established customers' portfolios. The reactions of customers to changes in activity flows are decisive in determining the outcomes for these firms. The interplay between rational, administrative supply chain management and the entrepreneurial management of a small sub-contractor is a core issue in this perspective. To the sub-contractor, such a perspective offers either the termination of relationships or being subordinate to a strong administrative power. Administrative co-ordination in the supply chain may become so strong that small sub-contractors' positions and performance are threatened. Then the countervailing force to administrative stiffness, which Perrow (1992) sees as entrepreneurial action (involving a small firm's integration of conception and execution), may be damaged. In this case, the result may be a paradoxical return to the Chandlerian integrated organisation. Only the container is different.

The network based perspective

In the network-based perspective, competition and collaboration are seen as systemic phenomena. Small enterprises - here sub-contractors - are seen as being organised into localised small firm networks and viewed as being embedded in a social and institutional context. Different theoretical perspectives co-exist on this issue (e.g. Nygaard 1998 or Isaksen 1994). Alternatively, these small enterprises are seen as being positioned within vertically organised supply chains.

Contributions to this field seldom regard sub-contractors as operating in both local networks and vertical supply chains at the same time.[34] But in reality, they do. Operating in both worlds results in corporate based networks with strong administrative rationales being confronted with the traditions and norms of craft production, together with the informal rules and practices of co-ordination, socially embedded and exercised in localised business clusters.

Because co-ordination mechanisms are so different between these two types of network, it might be expected that interactive outcomes would be unpredictable. Just as organisations are transformed when they collaborate, the configuration and functionality of networks is transformed when actors with different values, traditions and behaviours meet in the network, and when the network interacts with large organisations and other networks. Some small business industrial districts and clusters are, in a systemic way, dependent on a few large enterprises. This, for example, is the case in the Danish Stainless Steel District, where three TNCs compete in the dairy equipment business (Christensen and Philipsen 1997). However, TNCs are also sometimes heavily dependent on the interactive efficiency of industrial clusters. This aspect of dependency shows up when TNCs try to source in low-price areas, emphasising the fact that low price is not the same as low cost.

A combined perspective

There is, however, no reason to believe that only one coherent network ever governs a particular supply chain. When transnational contractors configure their supply chains they have to link up and cope with local supply networks and differences in national production systems. This is actually one of the essential traits that provide global supply chains with versatility, i.e. it is here that the basic flexibility of the global supply system created.

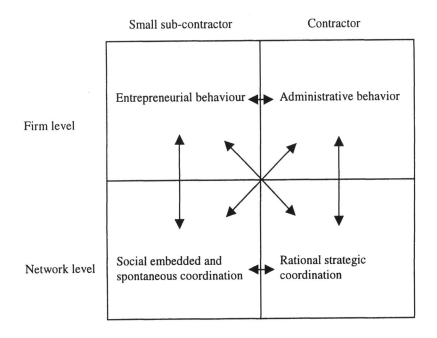

Figure 4.4   **Analytical levels and perspectives on the tensions between entrepreneurial and administrative behaviour in supply chains**

However, it is also important to recognise that supply systems with similar technological features may be organised quite differently, reflecting the fact that the governance structures of particular input-output arrangements have many layers and involve a number of strategic actors. Storper and Harrison (1991) have tried to model different types of supply chains in the context of industrial districts.

In Figure 4.4, an attempt is made to represent the interplay between typical sub-contractors and contractors and typical networks. The typology is based on Perrow (1992) and the distinction he made between stand-alone SMEs and a small firm networks

Beginning with small sub-contractors, a distinction can be drawn between sub-contractors that stand-alone and sub-contractors embedded in localised clusters of related enterprises. In the first case, the small entrepreneurial sub-contractor is confronted by an administratively orientated contractor based in a corporate network and a rational (strategic) culture. In this case, following Perrow (1992), the gap between conception and execution is increasing, implying that the entrepreneurial element in the chain is being repressed. In the case of the small sub-contractor embedded in a network of related co-specialists, the picture is different. Here, the small sub-contractor is embedded in a network of spontaneous co-ordination, implying that alternative business opportunities are open, although this depends on the degree of customer dominance. The key point, however, is that the administratively orientated contractor must use far more power to produce a gap between conception and execution, implying that an embedded entrepreneurial pattern is more resistant to supply chain pressure than in the stand-alone sub-contractor.

From a slightly different angle, an interface for mutual learning can be identified, both at the firm level and at the cross-functional level, i.e. between entrepreneurial behaviour and rational strategic co-ordination. From this perspective it is important to note that the contractor will try, and will sometimes succeed, in codifying 'sticky' knowledge, thus making it mobile. This point has important implications for localised learning systems, namely that task partitioning often leads to the separation of specific key functions that may very well become mobile. Again this situation can be exemplified by the case of the Stainless Steel District in Denmark (Christensen and Philipsen 1997). At the outset, construction skills in this District were embedded in enterprises making components or the sub-assemblies. Over the years, this competence turned into professional engineering skills. This engineering competence then accumulated in a small number of separate engineering enterprises. These enterprises are now vulnerable to take-over, and one of the engineering offices owned by a TNC was recently relocated.

There are also many examples of buffering arrangements in supply chains. In these situations, buffer units have been installed as a cross cultural unit to mediate between administratively driven transnational contractors and small firm clusters. The buffer might be a local production unit that has been acquired by a TNC.[35] In such cases the first generation of managers balance informal patterns of co-ordination institutionalised among the small sub-contractors in the area with the administrative claims of the new corporate owners. Experience from Denmark indicates that problems in balancing these diverse interests and understanding the spirit of the small business district arise when first generation managers from the area are succeeded by second generation managers who lack the trust of the small businesses in the area and do not identify with them.

Concluding remarks

In this chapter the focus has been, on the one hand, on the positions of small sub-contractors in localised clusters and communities and, on the one hand, on their positions in global supply systems. It is a dual position with lots of inbuilt tensions and paradoxes. But, it is also a key to improving our understanding the dynamics of regional industrial clusters and global supply networks. Such an understanding is urgently needed, since most current research tends to compare rather than combine these two perspectives.

SME sub-contractors are seen as important elements explaining the dynamics of global supply systems. TNCs tap into these localised clusters and tend to impose their administrative cultures on them. This may, paradoxically, bring about the disintegration of these clusters, as has been the case in Japan. In turn, this 'globalisation' of supply patterns has tended to reduce geographical differences which, ultimately, harm or retard industrial dynamics. Small sub-contractors' commercial opportunities are intimately involved in these evolving global supply systems.

SMEs that have lost their dynamism are unable to cope with structural change. They are particularly vulnerable when parent firms, suffering from their own rigidities, cannot salvage themselves from downturns. The promotion of industrial networks may be an effective policy if it can be used to instil mutual learning, dynamism and renewal within and economy. But, equally, to rekindle that dynamism, over-reliance on networks themselves can lead to decline as a consequence of lock-in. The key to successful SME sector development, therefore, is not simply the promotion of networks, but the promotion of dynamic networks, which can effectively deal with economic upturns and downturns.

5 Industrial Change and Local Competitive Advantage: Industrial Production Systems in Turin

Paolo Giaccaria

Introduction

This chapter explores the impact of change and the shift from Fordist production to flexible post-Fordist systems in the place-specific context of engineering industries in Turin. For some years now the interrelationships between the size of an enterprise, competitiveness and territory have occupied a significant position in socio-economic analysis. In other chapters in this volume the context is outlined of the changes that have occurred in the world economy within which the dynamics of these relationship can be interpreted: deverticalisation of the production process, transformation of the supply of goods and components, the formation of global commodity chains, the growth of the role of small and medium sized enterprises, and the changing provision of finance and services.

The analysis in this chapter is concerned with the impact of these changes on the systems of manufacturing production in Turin and the subsequent emergence of new systems of production. In the first part of the chapter, the nature of competitive relationships between companies is discussed and the impacts that change have had on them. Change, it is argued, has transformed Fordist relations of production into either a market driven post-Fordist mesosystem, or a local value production system based on networking. The second part of the chapter examines the applicability of these two emerging models to the transformation of the manufacturing economy in Turin. The Fordist Fiat related vehicle production that dominated this district in Italy through to the 1970s and 1980s, is shown to have been transformed into at least three distinctive new systems, related to these two ideal type models, each of which offers Turin a very different economic future.

Company competitiveness and local competitiveness

The first concept that needs to be considered is that of 'competitiveness'. This can be defined, in very general terms, as the capacity of one agent to succeed against other competing agents, respecting the rules set by a third party. In the case of production, this means designing and selling a product, at the expense competing producers of similar goods, and making sufficient profit to maintain the competitive advantage.

The hegemony of internal relationships

How can competitiveness be achieved? In general terms, the activities of an economic agent are concerned with organising the relationships that make up and impinge upon its operations: relationships between production processes and investment, inputs and costs, outputs and revenues, and research and innovation to achieve productivity and profits.

A company that can efficiently organise these relationships is able to acquire and exploit competitive advantage. Depending on how economic agents organise these relations, they will obtain varying success in the market. From the orthodox economic perspective, organising to achieve efficiency is an internal company concern. It is up to the company to decide how to organise its production, what investments to make to increase the productivity, and what quality it seeks to achieve. Obviously, traditional economic theory does not deny the existence and importance of companies' external relationships, but it limits itself to assuming that these relationships with the environment (with other companies, the market, the workforce, and so on) are governed by decisions taken inside the company.

To conceptualise the dynamic between companies' internal and external relationships, use is traditionally made of two main concepts: *hierarchy* and *market*. This distinction was used by Williamson (1985) to explain a company's choice between performing functions internally or having them performed externally by other companies. The decision to 'make' implies the dominance of the hierarchy of internal organisation. The decision to 'buy' implies that a company externalises functions to the market.

In this chapter, however, the terms 'hierarchy' and 'market' are used somewhat differently. The focus here is not on the internal company decision to 'make' or 'buy'. Instead, 'hierarchy' and 'market' are interpreted as two ways of organising relations *outside* the company. In the case of the term 'market', the reference to external relationships identifies exchanges with other companies. In the market place each company possesses a single identity (seller or buyer) and establishes with other companies passing relationship that do not go beyond trading transactions irrespective of the identities and histories of the two parties. The term 'hierarchy' used in the context of companies' external relationships relates to dominance structures that 'lead' companies typically develop over subordinate suppliers in Fordist systems of production. The use by companies of outside suppliers, even if they are small and medium sized enterprises, is not in itself indicative of an industrial district form of organisation in a local economy. The creation of vast networks of suppliers is just as typically a variation of Fordism, as is evident in the automobile industry, for example (Conti and Enrietti 1995). However, the relationships between the 'lead' company and its suppliers are characterised by administrative, commercial and technological dependency. These are certainly market-based relationships, but they are also hierarchical, not because control is exercised through shareholdings, but because control is exercised just as effectively in other ways (see, for example, the discussion in Chapter 9).

To summarise, the terms 'hierarchy' and 'market' are used here to label different ways of organising relationships *between* companies. The first term highlights relationships of dependency: with a 'lead' company imposing its own identity on that of its suppliers. The second term is used to denote relationships that are purely trade based: where the identities of the parties involved in the exchange remains completely separate.

The emergence of new external relationships

Profound change, however, has undermined orthodox interpretations of competitiveness and the relationships between companies, as explained in the previous chapter. For the argument developed here, change now means that companies' internal relationships are no longer the sole determinants of their external relationships. A company can, in fact, organise its internal relations effectively without this having a beneficial effect on the organisation of its external relationships. For example, the fact that a company invests its profits in new machinery can bring it competitive advantage and, at the same time, damage relations with the workforce. In the same way, the adoption of new technology can bring about the bankruptcy of numerous suppliers and an impoverishment of the local production system on which the company itself might be dependent.

Specialisation and success in niche markets makes external relationships and geographical proximity increasingly important. It seems clear that what drives specialised companies to concentrate in specific areas are particular benefits that are difficult to control and reproduce within the individual company.

Indeed, the relationships that determine competitiveness are increasingly externalised instead of being managed within the company. Now, production processes involve an increasing numbers of specialised suppliers, many of whom, because of their size and innovative capacity, are as important as the lead companies themselves. The innovation process is no longer contained exclusively within the confines of the company, but is triggered by interaction between the various actors in the same locality - suppliers, customers, public and private research centres, designers, consultants and so on. As far as labour is concerned, relationships between the technicians and experts of customer and supplier companies are of ever increasing importance. It is, in fact, through informal communication between people who share the same technical know-how and language that information vital for company competitiveness is transmitted. Markets too are becoming more differentiated, creating needs and issues not fully controllable from inside the company. Therefore, consortia and associations of companies are essential for success in complex global markets. As a consequence, investments are not only related to the acquisition of resources, but assume more and more the character of investments in relationships. Establishing relationships with prestigious customers or participating in innovation projects are no less important than making investments in new machinery. Both demand considerable inputs of human and financial resources.

A new role for territories

Two conclusions can be drawn from the discussion of the previous sections. First, the external relationships on which the competitiveness of companies partly depends are now increasingly spatially circumscribed. They are not, in fact, immediately classifiable to the two traditional categories of 'hierarchy' and 'market', but imply a greater degree of complexity. To establish collaboration between a company and a research centre, or to develop the exchange of information between the technicians of companies linked as buyers and suppliers, it is necessary for those relationships to have characteristics that were once neglected in economic analyses. The characteristics include personal contact, trust, continuity, shared values, skills and language. These characteristics often depend on actors being located in the same territory. Proximity is important not simply to reduce transaction costs. More importantly, geographical proximity is an essential condition for creating a community of economic and social actors, founded on common customs and values, on the continuity of human relations and on understanding and trust. The importance of external relations is therefore translated into the importance of 'local' relations. Place and locality is where the economic community consolidates its own competitive advantage through daily practices. In this sense, competitiveness cannot be separated from *embeddedness* in place (Grabher 1993b).

Second, it is now possible to talk of the *competitiveness of territories*, and not just the competitiveness of individual companies. Separating the destiny of regions and companies has led numerous local administrations to set up renewal programmes for their manufacturing sectors to valorise local resources. Thus, terms like 'regional competition' and 'urban marketing' have gained currency in urban and regional planning. Reductions in wage rigidities, assistance in location decisions, the improvement of disused industrial sites, and the reduction of utilities costs, are only some of the measures implemented by local authorities to attract new investment to compensate for crises in existing industrial structures.

From this discussion, a distinctive logic emerges for interpreting the competitiveness of territories. In this, a territory is competitive to the extent that it possesses and supports a rich fabric of relationships between the economic actors on which the competitiveness of individual local companies depends. Contemporary factors of location are now principally relational and socio-cultural. They include manufacturing tradition, skills, trust and so on. And, they are not easily transferable between places. A territory that possesses, defends and develops these relational factors not only supports the competitiveness of local companies and sectors, but can also attract companies that have understood the need to *embed* themselves in order to gain local competitive advantage.

This change in focus from internal to external relationships, and from the individual company to territory, has important implications for research, some of which have been examined in detail in other chapters (see in particular Chapters 2 and 4). First, the growing complexity of inter-firm relationships that has accompanied change makes it necessary to adopt new descriptive models of economies. The development of network concepts is particularly important in this respect. This metaphor has allowed analyses to step beyond the dualism of hierarchy and the market which have been the traditional points of reference for understanding the organisation of social and economic relationships. Second, attention has been focused afresh on the economic role of social, cultural, environmental and political factors. It is increasingly clear that the economic realm is inseparable from the other dimensions of human agency. The economy co-evolves with the social institutions within which it is embedded. Third, this process of evolution has been expressed through the metaphor of learning. This concept suggests that the formation of knowledge and innovation depend on the joint communicative action of those local actors who *learn* to transform local resources into competitive advantages. Fourth, to understand this new relational economy means abandoning over-simplified contrasts (such as between Fordism and post-Fordism) in favour of schemes of interpretation that allow us to grasp the ambiguity and multiplicity of the contemporary economy. It is not, therefore, a question of uncritically contrasting old and new, TNCs and systems of small companies, but of understanding how different production models co-exist and co-evolve.

In the next part of this chapter an attempt is made to recognise new models of inter-firm relationships that take account of these recent transformations - the organisational innovations introduced by the growth of small and medium sized enterprises, and the new role of territory in fostering economic competitiveness. In particular, attention will be focused on the production of intermediate and producer goods, rather than on goods for final consumption.

Towards a model of global/local connection

From the preceding discussion it is clear that change, triggered by processes of globalisation and the drive to achieve increased economic flexibility, has significantly altered inter-company relationships and the bases of competition. Change has impacted on three aspects of these relationships:

- their spatiality – which can be either global or local:
- their nature – as input or output, or as buyer or supplier transactions; and
- their form – their organisation and orientation within markets, hierarchies and networks.

We can interpret the positions of economic operators actors in different systems of production according to their unique combinations of these characteristics. As a result, every economic actor (company, group of companies, sector or industry) can be conceptualised as occupying a unique position in the structures of productive relationships that prevail at any particular time and in any particular place. Change undermines the bases of competitive advantage that underpin these structures of relationships, altering and radically transforming them to produce new structures.

As change has removed the old certainties of relationships structured along Fordist lines, at least two new models or systems of inter-firm and inter-company relationships have emerged, or are in the process of emerging. The first is a post-Fordist mesosystem driven by market mechanisms and hierarchies of market dependency. The second is a local value production system (LPVS) based on network relationships between companies and powered by socially derived reciprocal relationships of trust and loyalty that stimulate local processes of learning. These new models of inter-company relationships are elaborated in the next section.

The post-Fordist mesosystem

The Fordist model of enterprise relationships divides companies into two sets, lead enterprises and dependent suppliers. The 'lead' enterprise is the core of this Fordist model. Irrespective of its form (oligopolist, monopolist, multinational or transnational), it has the capacity to control both its own organisation and that of companies in its surrounding environment. It is an actor that operates in a global decision-making context and its organisation is a blend of both 'network' and 'hierarchy' elements. The network element comes from the world-scale organisation of transnational corporations, which connects relatively independent local units into a corporate network. At the same time, relationships within that corporate network are fundamentally hierarchical. Indeed, the individual units and divisions are still orchestrated by a central management structure, which sets strategies, monitors performance and has the final say on their survival.

Dependent suppliers provide lead enterprises with inputs through strongly dependent and exclusive relationships. Traditionally, local suppliers are often left with little autonomy and competitiveness and are organised into a hierarchy under the control of the lead enterprise. Now, after the restructuring and globalisation of the Fordist economy, individual companies in this group have followed different courses. Some have grown and gone global, becoming the partners of lead enterprises. Some have been taken over by networked multinationals interested in being suppliers to the same lead enterprises. Still others have retained their roles or have been absorbed into the fabric of local sub-contracting systems. Nevertheless, what characterises this type of company is that a single local plant remains very much dependent on the strategies and organisation of a dominant enterprise (also see Chapter 6).

From the market viewpoint, a production system comprising a lead enterprise and its subordinate suppliers is still able to serve the customised mass markets that have evolved from Fordist mass markets. The system, however, no longer sells a single product on all the markets. Products, (consumer durable, intermediate and producer goods) are now increasingly customised to meet more diverse and sophisticated demand.

The new form of integration that is emerging, that combines local, global, network, hierarchical and market dimensions, is a meso-economic system or *mesosystem*. The identification of such a mesosystem stems from the need to recognise a level of activity intermediate between the micro- and the macro economy - an intermediate space where, within a given production system, its is possible to define:

> "... sub-groups of economic agents, with their own and/or highly-interactive coherent and long-term dynamics; strategic behaviours of organisations; structural economic conditions; and rules of the institutional game." (Gilly 1994, p. 296)

In the context of local development, identifying a mesosystem helps to explain the countless situations in which general solutions are not found only local ones. In some respects a mesosystem encompasses characteristics of local systems: the importance of regularity and durability in relationships; the role of territory; the importance of networking and external economies of scale; and the significance of history. When Gilly (1994) defines a production mesosystem as an "historically constituted organised system of long-term relations between production organisations, equipped with group production dynamics" (p.298), it is clear that it also embraces the behaviour of large networked multinational companies and global supply chains.

This first model of inter-firm relationships is labelled here as the *post-Fordist mesosystem*. It is a *mesosystem*, since its behaviour is not determined by the strategies of a single company but by the relationships between many economic actors. It is a collective entity intermediate between the individual company and the global economy as a whole. However, it is not a *local* system since its identity is not defined in terms of territory and it is not solely dependent on local relationships. It is *post-Fordist*, because the system derives from the evolution of Fordist oligopoly and, in a certain sense, represents an organisational solution to the changes being experienced in contemporary market economies.

The local value production system

The second model is the local value production system (LVPS) in which the subject is not the company but the *company network*. The archetypal form of this system is the industrial district where a number of companies are linked by trust, mutual knowledge and know-how into a network of co-operation and competition. As a result, competitiveness is not just an attribute of the individual company, it is an attribute of the network of companies as a whole, with its own local identity.

The market for this system comprises two components. First there are *strong relationships with the local market*. This market is among customers belonging to the same local system rather than being with a lead enterprise. Only later will the production network be able to develop relations with wider and global markets. This second market component develops from international demand for highly customised products. Customisation, in this context, shields small producers from competition with large ologopolistic producers. The relationships that are created with these global customers go beyond mere market relations. They also involve trust, and knowledge of mutual requirements, capacities and traditions.

There are many examples of these networked systems of production and marketing, and their international significance appears to be increasing. However, they all share a set of common characteristics as *local value production systems*. First, the relationships on which they are built constitute a *system* because they incorporate all the relationships that bind local companies together, not just the functional ties of buying and selling. Those relationships also include links for the circulation of innovation and knowledge. Second, the systems are *local* because they are built on place-specific resources that are not transferable which makes them internationally competitive. Third, *production* in these systems refers to more than just output. It also refers to other dynamic aspects of firms' activities including innovation, firm formation, job creation and exports, for example. Fourth, the basis of the system is to *add value* in production. But here value must be interpreted in the broadest possible terms, referring not only to the concept of added production value - and, therefore, to productivity - but also to the added 'social' value that arises as employment, knowledge, innovation, growth and remuneration.

In the second part of this chapter these models of post-Fordist mesosystem relationships and the relationships of local value production systems are explored in the empirical context of the restructuring of manufacturing in Turin. Through this analysis it is possible to move from abstract generalisation to the concrete detail of change as it is experienced locally.

A new image of Turin's manufacturing system

In the 20th century, the presence of Fiat and its capacity to organise and profoundly structure territory gave Turin its most well known image. Over time, the identity of Turin was condensed to that of a car-producing city, the city of Fiat. The name Turin evokes Dickensian scenes and arouses comparisons with the grey centres of early industrialisation, such as Manchester, Liverpool or Lyon.

Since the 1980s, however, the relationship between Turin and its Fordist past have changed radically in three ways. First, as production has been increasingly relocated to southern Italy, Turin's vertically integrated production system has been reduced to the production of only the highest valued components and the assembly of the most prestigious of Fiat's models. Second, the relationships between Fiat and its suppliers have changed radically. In the 1980s and 1990s, the numbers of first tier suppliers were drastically reduced, with surviving suppliers gaining power through increased participation in innovation and component design.

Third, change in the automotive sector has made the greater integration of research, innovation, design, engineering and production a necessity. The designer combines the results of research into materials, aerodynamics, safety and environmental issues into prestigious and innovative stylistic solutions, fundamental for competitiveness. Design ensures that product and process innovation can be used in production.

To examine the transformation of Turin's manufacturing structure, a questionnaire survey of engineering companies was undertaken. The survey covered 150 companies operating in five local production systems in the province of Turin - Torino, Avigliana, Pinerolo, Ivrea and Rivarolo. The engineering companies were, producers of:

• *producer goods* (44 interviews), including makers of machine tools and measuring machines, but also companies engaged in related activities such as the design and manufacture of industrial plant, and the design of integrated production systems;
• *vehicles* (62 interviews), including specialist vehicle components producers and also producers of boats; and
• *non-specialist engineering goods* (44 interviews), some of whom have historical and geographical ties to car production, but who work for numerous firms in other sectors including household appliance

manufacturers, aerospace companies, and machinery producers and companies making finished goods such as locks and handles.

The questions focused on:

* *local/global relations* - especially the internationalisation of the local manufacturing structure;
* *inter-company relationships* - the continuing existence of hierarchical forms of organisation (particularly among multinational groups) and the formation of non-hierarchically organised groups of companies; and
* *learning processes* – and the increasing importance of Turin's designers and engineers and their relationships with engineering companies.

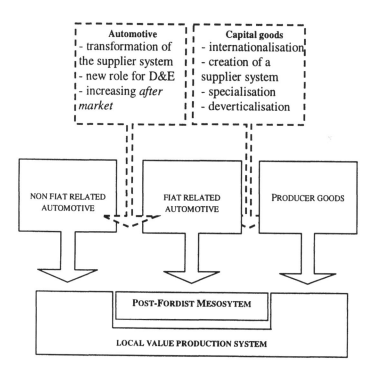

Figure 5.1   The transformation of production systems

The companies that were interviewed were asked their opinion on the role of traditional local factors in determining competitive advantage: logistics and accessibility; the presence of other local agents; training; relations with the workforce; and institutional support. They were also asked their opinion on the local business environment and local 'industrial atmosphere'.

The empirical analysis highlights how change in Turin's engineering industries is part of a more radical reorganisation of the district's production. The district now comprises three distinct systems: *a producer goods system, a vehicle production system dependent on Fiat*, and *a vehicle system independent of Fiat* (Figure 5.1). Each of these systems has its own distinctive form of organisation that goes beyond Fordism and assumes the characteristics of the previously described concepts of the Post-Fordist Mesosystem and the Local Value Production System (LPVS). The first of these concepts describes the 'Fiat galaxy' while the second encompasses the transformations that have occurred in both the producer goods system and the vehicle and components production system that is independent of Fiat.

The post-Fordist mesosystem

The empirical analysis suggests that the relationships of the Fiat mesosystem have weakened, questioning the traditional assumption that Turin's manufacturing base is exclusively orientated to automotive production in general and to Fiat in particular. The hierarchical production system that remains is, however, typical of Fordist Turin.

The competitiveness of the mesosystem still depends on companies having preferential relationships with major corporations, principally Fiat. It is true, nevertheless, that change has progressively empowered first tier suppliers. However, these relationships remain profoundly hierarchical, with decision-making and innovation flowing through various levels, from Fiat at the top to small and medium sized sub-contracting companies at the bottom.

This continuity of Fordist organisation is particularly evident in four aspects of current relationships. First, the relationships between companies show the emergence of network characteristics additional to persistent hierarchical features. On the one hand, this involves the transformation of the Fordist multinational, Fiat, into a networked transnational corporation (TNC). On the other hand, there are also signs that innovation is spreading to small enterprises as networks centred on Fiat develop to embrace design and engineering studios (D&E), first tier suppliers, and Fiat Research Centre programmes. Nevertheless, dependency on Fiat remains. Second, hierarchical relationships still frame local and global interactions. Either Fiat or first tier suppliers mediate access for firms to international markets. But, the survey responses show that international groups that act as nodes in both local and supra-national networks are not embedded at both scales. Progressively, they are becoming less locally embedded as their exports expand. Indeed, the rise of national and international groups appears to be diminishing territorial embeddedness without bringing the benefits of connection to global markets. Third, the learning relations linked to design and engineering work are still being channelled primarily through companies' main customer - Fiat. In this mesosystem, the use of D&E for product and process innovation is less frequent than in the system making producer goods. Perversely, while components manufacturers linked to Fiat make greater use of D&E, many of them have no design and engineering activities and depend entirely on those of their customers. In this hierarchical decision-making chain, knowledge and information is in the hands of the dominant firms. Fourth, the final product of the mesosystem, the automobile, is being sold on the customised mass market that has evolved from the Fordist mass market.

From this analysis of milieu relationships, it is reasonable to conclude that despite continuing local bonds, the Fiat supplier system is structured as an a-territorial economic mesosystem. The organisation of the Fiat mesosystem appears to be in a process of transition from old Fordist arrangements to a post-Fordist identity that is still to be defined, but which is somewhere between the global and the local and between hierarchy and network.

The emerging post-Fordist mesosystem is made up of various actors each following different trajectories of change and each of which will contribute differently to the system in the future. At the centre of the system is Fiat, which is becoming increasingly transnational, and assuming a global network form. Its territorial embeddedness in Turin is increasingly ambiguous and controversial. On the one hand, it is part of a network of local co-operation, especially for decision-making, innovation and engineering activities, because the designers and first tier suppliers it uses are strongly locally embedded in Turin. On the other hand, the locations of production plants are now seen in a global perspective, with little concern for Turin.

First tier suppliers appear increasingly to be on the borderline between hierarchy and network forms of organisation - between local embeddedness and a-territorial dependency. However, the inter-firm relationships of these enterprises are still evolving. Their progressive involvement in co-design networks with Fiat has increased their autonomy and facilitated the consolidation of network relations. But, their acquisition and take-over by multinational groups has meant that many first tier suppliers have become part of a new hierarchy. Ambiguity can also be seen in their evolving relationships with the local area. While these networks embed companies in a close-knit fabric of local non-trade relations, it is still true that many of these companies are willing to relocate their manufacturing activities close to the new globally distributed plants of their main customer (also see chapter 6).

Sub-contracting is the component of the system that most closely reproduces the characteristics of the old Fordist form of organisation. Sub-contractors remain largely excluded from the processes of restructuring of the Fordist hierarchy and are unaware of the birth of new relational networks with either first tier suppliers or other sub-contractors. From a territorial point of view, sub-contracting acts within a mainly local context in contrast to the findings reported in Chapter 4. These small firms can gain access neither to global production and innovation networks nor to foreign markets.

Finally, there are two other important sets of actors that 'network' with Fiat and its first level suppliers. These are the design and engineering companies and Comau, the group controlled by Fiat that makes producer goods for the holding company (especially robots and integrated production systems). In both cases, these are actors whose competitiveness and excellence is based on local factors which cannot easily be found outside the Turin area (tradition, know-how, experience, skilled personnel, trusted suppliers).

The image that emerges from this analysis is that of a system that is shifting towards the globalisation of the relations of production and, therefore, towards potential uprooting from the Turin milieu. The rooting of the mesosystem in Turin depends, in fact, on the balancing of two opposing forces:

- a *centrifugal* force, resulting from the strategic decisions of global actors – the buying up of local suppliers by outside interests, and Fiat's global decentralisation that encourages its suppliers to follow; and
- a *centripetal* force, linked to specific localised cognitive processes, the most important of which is the close connection between D&E activities and production.

A balance between these forces has yet to emerge.

The local value production system

The empirical analysis also underlines the emergence of another local system characterised by a stratification of local and global relationships which sets it apart from the previously discussed system. This is the local value production system (LVPS). The LVPS identifies those activities and sectors which, starting from the engineering tradition that is the heart of Turin's know-how, have been able to activate the local factors (trust, personal acquaintance, sharing of values and skills etc) that play a fundamental role in supporting the competitiveness of local manufacturing systems. At present, two variants of the LVPS can be identified in Turin, a producer goods system, and a non-Fiat-related automotive system (Figure 5.1).

The heart of the LVPS is a nucleus of relationships that link design and engineering (D&E), the manufacture of producer goods, and a close-knit network of companies specialised in micro-mechanics and information technology (IT). This is the first and principal element of the LVPS. The competitiveness of this system arises from its propensity to export. The member firms tend to have large numbers of customers, rather than one or two key ones. They are highly specialised and globally market orientated. At the same time, they see the role of the local economic environment of Turin in a positive light in terms of 'industrial atmosphere', manufacturing tradition and so on.

The milieu of the producer goods system derives from three aspects of the interactions between companies that define the identity of the local manufacturing system. First, companies are involved in networks of informal relationships, especially with their customers. Products are tailored to the highly specialised needs of those customers who are mainly located abroad. As far as supplier relations are concerned, a distinction can be made between those with IT consultants (for software) and those with the micro-mechanics sector. Here quality is a major issue, and more so than in the vehicle sector. Second, local/global connections are radically different to those of the Fiat mesosystem. In the absence of major international groups, the link to the global network is through numerous small and medium sized companies that sell directly on international markets. Here, market access is not mediated through a few large actors. Most of the manufacturers and their suppliers of tools and components are export orientated. The informal and co-operative nature of their commercial relationships facilitates the transfer of skills and knowledge and provides access to innovation. Third, inter-firm relationships involve learning through customer-focused transactions centred on co-design and co-engineering. With customers located abroad, these are not entirely local learning processes, though firms in the LPV system make intensive use of the D&E skills in Turin. In part, territorially embedded learning processes persist.

Thus, in the design and engineering element of the producer goods form of LVPS, firms are strongly embedded in the local Turin context. They draw their competitiveness from the continuity of the Turin manufacturing tradition, on which they have been able to build advanced and innovative competencies in micro-mechanics (especially in aerospace production), mechatronics and IT.

A second element of the local value production system (LVPS) centres on vehicle production activities *not* linked to Fiat. Some firms have shifted to producing components for motorbikes and agricultural machinery, others to producing for the automotive spare market. The drivers of this transformation are medium sized and large, internationalised companies, some of whom were formerly first tier suppliers to Fiat, who have adopted diversification strategies. They have drawn around themselves significant parts of the Fiat supplier system, creating a new element within Turin's engineering system. From a territorial point of view, this engineering system has features intermediate to those of the Fiat-centred mesosystem and the producer goods system:

• relationships between companies remain fundamentally hierarchical;
• firms make greater use of information external to the local system, depend less on one main customer, are more market orientated, export more, are more dependent on local logistics support and have fewer corporate groups in their ranks; and
• design and engineering activities are in a phase of incipient diffusion amongst the companies in the system.

These characteristics make this particular vehicle segment (not linked to Fiat) one of the critical kernels of local development in Turin. The manufacturers of the producer goods system and the Fiat mesosystem possess their own clear development trajectories; the former based on embeddedness and the latter on globalisation. For the vehicle segment not linked to Fiat the situation is more critical. It appears to be in a delicate phase of transition in which a supportive network of personal and entrepreneurial relationships has not yet been formed. At the same time, the globalisation of the automotive sector, and the desire of local entrepreneurs not to belong to the Fiat supply system, are weakening its ties with the Fiat mesosystem.

Thus, the local value production system (LPVS) has two components. One comprises companies that constitute the producer goods system, and the other comprises the firms of the non-Fiat vehicle system. In the first, businesses have a very local production perspective and are linked into networks based on trust and the sharing of specialist skills. They appear to be strongly embedded in the territory from which they draw the resources needed to maintain their own competitiveness in international markets. In the second, involved in vehicle production but independently of Fiat, only some businesses in the system are strongly locally embedded. Others are not. This is an unstable component of the LPVS that is experiencing profound transformation.

Overall, the markets for Turin's local value production systems are *highly specialised, personalised* and *internationalised*. They are global markets in vehicle spare parts, in the non-Fiat vehicle sector, and in niche markets for both intermediate and consumer goods. Network relationships linking producers and consumers in these markets are fundamental for the maintenance of competitiveness.

Nevertheless, it is clear that local markets still remain important for local value production systems. The Fiat group and its suppliers have played a dual role in this respect. While, on the one hand, they themselves have created a large pool of demand, on the other hand, they have also functioned as a 'technological incubator' in the sense that many new entrepreneurs are technicians or workers who have left the Fiat mesosystem.

Obviously, the local value production system and the Fiat mesosystem are not reciprocally closed systems. In addition to the market relations already mentioned, there are various other points of contact between the two. First, in recent years, Comau has progressively reduced its exclusive bond with Fiat, establishing itself as one of the world leaders in the design and manufacture of integrated production systems. Parallel to this, it has intensified its non-trade relationships with local and global company networks. This transformation makes Comau a potentially important catalyst of knowledge creation and innovation in the Turin area, generating positive externalities for the local value production system. Second, the design and engineering studios, although based on skills profoundly rooted in the Fiat mesosystem, are also important actors in the local value production system. Third, some first tier suppliers are gradually breaking free from Fiat and many small sub-contractors who formerly worked exclusively for Fiat are now sometimes accepting contracts from these companies.

Conclusion

The purpose of this chapter has been to examine the impact of change on the organisation of buyer-supplier relationships among the firms in the engineering industries of Turin, as they have sought to both maintain and create competitive advantage. Change has undermined the old certainties of Fordist production systems like the Fiat-dominated mass production system that has characterised the engineering industries of Turin for many decades. The Fordist style of organisation, in which hierarchical dominance structures linked Fiat and its suppliers, is now progressively being replaced. New, post-Fordist systems of production are emerging, involving networked relationships on both global and local scales. They are moving Turin's economy is a variety of directions towards an uncertain future.

It is clear is that the legacy of Fordism will not disappear in the short-term in Turin. Fiat's Fordist system is being progressively transformed into a mesosystem based on global networks of buyer-supplier relationships. Now, in addition, at least two local value production systems are emerging in Turin, built on the networked local embeddedness of firms. One system is concerned with the manufacture of producer goods, the other with those sections of the vehicle industry that are independent of Fiat. Both have their roots in the cognitive and social environment of the district's Fordist past. What is emerging, it would seem, is not a radical divide between Fordism and post-Fordism, but the tentative exploration of a range of new forms of organisation in an effort to gauge what will best maintain and create new competitive advantage in dynamic, uncertain markets.

6 Sub-Contractors, Supplier Parks and Supply Chain Management: The Case of Volvo's Arendal Supplier Park

Claes G. Alvstam and Anders Larsson

Introduction

During the last decade there has been a clear restructuring trend among suppliers of parts and components in the automotive industry that has produced a hierarchical spatial pattern consisting of a small number of first tier, or direct, suppliers complemented by several layers of sub-suppliers. The pressure for these changes has come from the large final assembly firms, as they have shifted responsibility for production, inventory and R&D upstream in the production chain. Such change in the organisation of production allows final assemblers to concentrate their resources on core activities, including marketing and the co-ordination of the supply chain.

Parallel and interrelated aspects of this organisational restructuring are the direct and indirect spatial impacts of just-in-time (JIT) production requirements. The aim of JIT is to achieve minimum inventory levels, short through-put times, zero-defects quality, and consequently higher levels of reliability and frequency in deliveries. Increasingly in the world automotive industry, sub-contractors are branch plants of larger international corporations that specialise in parts and components manufacturing. The 'classical' sub-contractors, the wholly independent small or medium sized enterprises, are losing ground or are becoming integrated within larger structures. Functionally though, these branch plants operate as independent business units with separate management teams. Their parent companies' requirements for profits and returns on capital may be, in fact, far more stringent than in family owned SMEs. The trend of creating supplier parks to accommodate the branch plants of global parts and components manufacturers (for example, the American Lear Corporation and the Norwegian company Hydro-Raufoss) may have been enhanced in the case of the Volvo Car Corporation following the company's acquisition by Ford in early 1999. That take-over has heralded the gradual integration of Volvo within Ford's global network of sub-contract suppliers. Indeed, since the empirical work for this study was completed, Lear has decided to relocate its production from the remote municipality of Bengtsfors, north of Göteborg, to concentrate its Sweden-based activities in the Göteborg area.

In this chapter we discuss the role of small free-standing sub-contractors in such a production system, in terms of production and the creation of a new supplier park, using the case of the Göteborg final assembly plant of the Volvo Car Corporation and its domestic suppliers. The impact of putting pressure on first tier suppliers to locate close to the final assembly plant is first discussed in theoretical terms of forces of concentration and dispersion operating on different tiers of the production hierarchy. From this, five main issues concerning supply chains, sub-contractors and supplier parks will be discussed:

• The role of small and medium sized branch plants and independent SMEs in the ongoing restructuring process within the automotive industry. What are the prospects for competition with larger firms for first tier contracts?

• It is in many cases possible to observe that small domestic production units are incorporated in large international supplier 'conglomerates'. How does such a pattern affect power relationships with the principal firm? It might be that a relatively small automotive company, such as Volvo, is faced with co-operation with a multi-client supplier belonging to a much larger firm. Accordingly, how will the trend towards the creation of global suppliers to the automotive industry affect the spatial pattern of the supply system? That pattern has generally been dictated by the final assemblers, but the question now arises as to what the consequences will be for specialised sub-contractors with far less negotiation power with their clients.

• To what extent will Volvo's supplier park contain small and medium sized branch plants of larger international manufacturers of parts, components and complete modules, as well as wholly independent SMEs? Also, what types of operations will they be involved in? Will their geographical proximity to Volvo's final assembly plant open opportunities for closer co-operation in R&D activities, or does proximity only give advantages in terms of physical delivery?

- What are the conditions for the propagation of knowledge and the creation of an innovative milieu within a supplier park?
- Is it possible to create local supply networks among new, greenfield establishments in a supplier park?

Theoretical framework

An obligational model of sub-contracting

A large number of authors have argued that industries with assembly-type operations are much more concerned with supplier relations today than ever before. Imrie and Morris (1992) emphasise qualitative factors over price as the rationale behind sub-contracting. The main reason behind this trend is the growing demand for customer responsiveness, consistent high-quality production and flexible deliveries. The extensive MIT study of the automotive industry (Womack, Jones et al 1990) identified five major tendencies in the restructuring of the supply system in the automotive industry:

- reduction in the number of suppliers to each assembly plant;
- 'out-sourcing' – the greater use of outside firms as suppliers;
- changing attitude towards quality-issues;
- more emphasis on information sharing; and
- more frequent deliveries.

Lamming (1993) made one of the most comprehensive empirical studies of the restructuring process in automotive supply systems within the framework of the International Motor Vehicle Programme. He interviewed 129 companies around the world in search of factors that have characterised the development in the automotive supply system since the early 1970s. An outcome of the study was a four-phase model of buyer-supplier relations within which current developments revolve around issues of partnership. The cornerstones of this partnership model are long-term co-operation, trust and interrelatedness.

The importance of co-operation and trust in transactions has been the focus of the work of the 'network-school' in business economics and organisational studies (Axelsson and Easton 1992, Johanson and Mattsson 1991). Their view of industrial relationships stresses the interrelatedness between actors in a network and, consequently, the importance of social forces in forming inter-firm links (see Burmeister and Colletis-Wahl 1997).

Organisation – from market relations to integration?

Compared to the traditional model of arm's-length market transactions between independent firms, the partnership model relies on mutual trust and, ultimately, on mutual economic benefit. Figure 6.1 presents the range of buyer-supplier relationships available to firms, ranging from market-based transactions at one extreme to vertical integration at the other. These alternatives should not be interpreted as discrete steps but rather as a typology for understanding a process. In reality, there are a large number of different and overlapping arrangements. The characteristic feature of the new types of buyer-supplier relationships from a geographical point of view is the need to combine flexibility and stability for both the principal and the sub-contractor. There is no one solution to balancing these needs either in the short-term or in the long-term. The outcome for any particular relationship is most likely to be a result of accumulated ad hoc decisions rather than the outcome of a logical sequence of management decisions.

All the forms of integration in Figure 6.1 can be found in an automotive supplier system, and the processes operating within it appear to be leading to the formation of large, often multinational automotive system suppliers. The establishment of these large suppliers is predominantly a result of the two processes indicated by the arrows in Figure 6.1:

• the restructuring of the traditional supply system through mergers and/or acquisitions of smaller firms by large automotive supply companies with major capacity both in R&D and production; and
• the out-sourcing of formerly internalised production, either through co-operative arrangements or by selling parts production to outside firms.

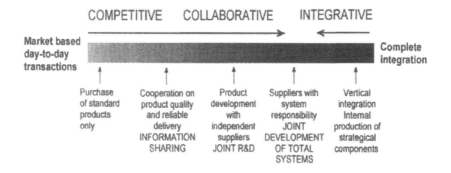

Source: European Commission 1996, Stralkowski, Klemm et al 1988

Figure 6.1   Stages of co-operation in a buyer-supplier relationship with respect to integration and commitment

The structural result of these organisational processes is the creation of a hierarchy of suppliers within a sub-contracting network. In the upper layers of the hierarchy are a small number of firms who become responsible for producing a particular system or module of components. To deliver these relatively complex products, there is a need for close interaction between the buyer and the supplier.

Supplier models in different regions

Japan

According to Sheard (1983), there is a distinctive organisation of sub-contractors in the Japanese automotive industry. Firms are arranged in a hierarchy with different layers of sub-contractors. The upper layer consists of large sub-contractors, not infrequently controlled by the principal contractors who are also their parents. These systems are a product of Japan's historical economic development. During the 1950s and 1960s, when the automotive industry was expanding, the systems were built from scratch using a pool of small and medium sized engineering firms, as described by Aoyama in Chapter 3.

Since then, these arrangements have developed into a system of tight inter-firm relationships between first tier sub-contractors and parent firms, often resulting in the establishment of formalised sub-contractor associations (Takeuchi 1990). The big first tier sub-contractors are often fully or partly owned by the large assembly firms who, in return for extreme flexibility, give long-term contracts and collaborate closely in product and process development. Typical of this Japanese system is the large number of small and medium sized firms in the lower layers of the hierarchy (Sheard 1983). These smaller firms have no contracts with the assemblers. They are usually engaged in the least complex work, and offer lower wages than upper tier firms. At the same time, Japanese industrial policy has promoted various forms of inter-firm alliance and collaboration, as was described in Chapter 3.

More recent studies of Japanese automotive buyer-supplier relations confirm this view, although there is evidence that the rigid company-centred hierarchies are starting to open up. This has been caused partly by the search for new markets and new technology, and partly by trade policy and the globalisation of supply linkages (Fujimoto 1997, Kalsaas 1998).

It is clear that the theoretical 'partnership-models' of buyer-supplier relations have found much of their inspiration in the Japanese example. It is important to point to the fact that behind the generalised picture of the Japanese model, we can find different forms of supplier organisation not always typically 'Japanese'. Within the hierarchical model, Toyota and Nissan, for example, have different histories and solutions in terms of supplier selection and management (Hayter 1997, Kalsaas 1998).

North America

The North American automotive sub-contracting system has developed in a different historical context and with different organisational outcomes. The most significant differences between Japan and the US systems are the long tradition in the US of vertical integration and market-based sub-contracting relationships (Rubenstein 1988, Hill 1989). Japanese firms produce about 30 percent of the value of a car in-house, whereas in the US the figure is approximately 70 percent (Hill 1989, p. 466). Hierarchical systems of sub-contracting are much less developed in the US than in Japan.

In terms of power relations, US sub-contractors are more usually independent firms – a contrast with the strong ownership ties in the Japanese system. Also, production for the automotive 'after-market' (i.e. spare parts, maintenance etc) and non-automotive markets is important for many large US sub-contractors, a situation that further minimises the power of the large vehicle assemblers (Glasmeier and McCluskey 1987). This tendency towards independence is also evident in the strategies of the large

US assemblers to separate their in-house components production into new companies; for example, Ford's components division as Viseton, and the GM components group as Delphi (Financial Times 1998).

Finally, it would appear that hardened competition and rising R&D costs have put US firms under pressure to adopt the Japanese model (Helper 1991, Glasmeier and McCluskey 1987, Mair 1994). This has the potential to expand sub-contracting and lead to the formation of sub-contracting hierarchies. But, there are significant social and historical differences in the organisation of production in the US and Japan which are evident in the Japanese 'transplants' that have been set up in the US (Mair et al 1988). These plants, which are a translation of Japanese production plants into the North American context, have been forced to adopt a combination of Japanese, tightly controlled, non-unionised arrangements and more independent US arrangements (Mair 1994).

Western Europe

The conditions under which the Western European automotive industry has developed are a different story again. Here there is a long history of parallel national production systems. Consequently, each national automotive producer has developed a home country base of domestic suppliers.

However, in the face of increasing global competition and the creation of the single European market, these conditions have changed dramatically. Now, the Europeanisation of supply systems is occurring, initiated by restructuring and concentration in the sector (Sadler 1997, Hudson and Schamp 1995). Also, the major ownership changes that have occurred in the automotive components sector globally, have begun to involve Western Europe, especially in the last five years (Financial Times 1996 a, b, c, 1998).

Evidence for Europe suggests that partnership models of buyer-supplier relationships have predominantly positive impacts on firm performance (Schamp 1995, Pulignano 1997, Laigle 1997). The implementation of these processes in Europe, however, appears to occur only with the introduction of new vehicle models or when greenfield investments are made – when there are opportunities to reconstruct working practices and supplier relations that tend not to occur in existing plants.

The Volvo experience

Historical development

The experience of the Volvo Car Corporation is unique. This is related to the historical trajectory of Volvo as a company, the concentration of its production in Scandinavia, the context of the Swedish economy and Swedish labour relations.

Volvo started production in Sweden in the late 1920s using independent and competing firms as suppliers. The final assembly plant was located at Göteborg and most strategic components such as engines, bodies and gearboxes were bought from external suppliers. These arrangements changed in the 1950s and 1960s when Volvo acquired most of its main suppliers, and started to produce components itself (Ellegård 1983).

In the early 1960s, the decision was taken to establish a final assembly plant in Ghent, Belgium. The main reason for moving abroad was the fear of staying outside the newly created customs union of the EEC. Accordingly, the organisation of capacity in Belgium was followed by the search for local sub-contractors to match the Swedish firms. During the first ten years, the number of sub-contractors increased in step with the expansion of production capacity at both locations, but between 1972 and 1982, the Göteborg plant stagnated while assembly in Ghent expanded. Moreover, in 1973 Volvo acquired a majority stake in Dutch DAF and started producing small passenger cars at that site (Elsässer 1995).

Both of these shifts in production capacity from Sweden to continental Europe contributed to a major change of Volvo's policy towards many domestic sub-contractors, who were faced with the requirement to follow Volvo abroad in order to retain their contracts with the Göteborg assembly plant.

The proposed merger between Volvo and Renault, that began to take effect during 1989-1993, also forced change on the sub-contracting system, brought about by the need co-ordinate Volvo's long-term product development with its larger, state-owned 'fiancée'. Even today, almost five years after the sudden 'divorce' of these companies, it is still possible to trace several elements in Volvo's current supply system to that Renault link-up. Similarly, the joint development venture that started in 1991 with Mitsubishi, at the NedCar Corporation in the former DAF plant in the Netherlands, can be seen to have affected Volvo's supplier policy for the S40/V40 model generation.

The board of the Volvo group signalled a further major strategic change in 1997, by stating its long-term intention to concentrate on heavy vehicles and construction equipment. These are areas of business in which Volvo has a large global market share, unlike the passenger car business. This announcement heralded a year of intense speculation on various proposed mergers, acquisitions and alliances, including potential partners such as Fiat, Volkswagen and Mitsubishi. It ended with Ford's acquisition of the passenger car corporation in January 1999.

The last 40 years have seen major changes in Volvo's work organisation and sub-contracting system with the company now concentrating on core activities and moving out of passenger car production. Many formerly Volvo-owned suppliers have been sold and the remaining activities have been reorganised into independent companies, such as Volvo Components Corporation and Volvo Transport Corporation (VTC).

Sub-contracting is not a cheap supply option for Volvo in Sweden because there are only marginal differences in labour costs between assembly-line workers and suppliers in the Swedish labour market. This means that cost saving as a rationale for sub-contracting has been less important for Volvo than for other vehicle producers. The fact that Scandinavia is only a small market for suppliers with competencies in all areas of the automotive industry has forced companies to search for suppliers on a European scale from the outset. This has produced widespread experience in international sourcing - experience that has been vital in the current phase of restructuring.

The structure of the sub-contracting system of the Volvo Torslandaverken assembly plant in Göteborg

Volvo's restructuring since the 1950s has exposed Sweden's domestic supply system to strong international competition. Already in the early 1960s, the Volvo Göteborg plant had almost half of its suppliers outside Sweden (Table 6.1). Non-domestic suppliers still predominate, with only 43 percent of supplier companies being located in Sweden in 1997 (Table 6.1). Expressed as the value of purchase, this level of foreign domination would probably be much greater.

Notwithstanding the changes in absolute figures, Table 6.1 shows sub-contractors from Germany, Belgium, France and the Netherlands have gained in relative importance. The most extensive losses have been among sub-contractors from Sweden, UK and the US. The relative growth of German suppliers is matched by the fall in numbers of domestic sub-contractors, while the growth for other countries is linked to Volvo's production in Ghent (Belgium) and Born (Netherlands), as well as the remains of the Volvo-Renault merger project. The greatest change in the sub-contracting system during these years was in the *number* of first tier suppliers. In 1960, over 1500 companies delivered components to the Göteborg assembly plant (Törnqvist 1963). That number had fallen by 75 per cent to 350 in 1997.

Table 6.1   Geographical pattern of sub-contractors to Volvo assembly plants in Göteborg, 1960 and 1997

Country	1960 #*	%	1997 #	%	Change 60-97 #
Sweden	823	53	154	43	-669
Germany	217	14	91	25	-126
France	9	1	26	7	17
UK	224	14	25	7	-199
Belgium	4	0	16	4	12
US	190	12	11	3	-179
Netherlands	5	0	11	3	6
Norway	37	2	8	2	-29
Denmark	33	2	7	2	-26
Switzerland	10	1	4	1	-6
Italy	1	0	3	1	2
Finland	6	0	2	1	-4
Others	7	0	17	5	10
Total	1559	100	358	100	-1201

* Number of employees is used as the measurement of size.

Source: 1960 figures: Törnqvist 1963, data for 1997 compiled by the authors

    Törnqvist's (1963) study did not present figures on sales or firm size. These figures are available only for 1997 and are summarised in Table 6.2. The geographical distribution of sub-contractors is shown in Figure 6.2. Table 6.2 shows that the majority of domestic suppliers are relatively low tech producers of metal and plastic components. It is likely that in the future, in the face of current restructuring processes, these sub-contractors will experience problems because they will be unable to take responsibility for developing and producing component modules. If they want to continue in the automotive industry, they will probably have to act as second or third tier suppliers.

    The geographical pattern of domestic suppliers to the Volvo Torslanda plant (Figure 6.2) shows a concentration of very small suppliers in the Göteborg region. Other suppliers in the 1-19 employees segment are mainly agents or retail dealers not engaged in production. It has to be pointed out that the maps show the situation before the start of the S80 series in April 1998. With the new suppliers included, there would be an even larger concentration of small suppliers around Göteborg owing to the establishment of the new supplier park that will be discussed in the next section. The maps also show how proximity to Göteborg decreases with firm size. The exception is the 100-249 employee size category where southwestern Sweden is clearly dominant. This can be explained by the fact that within this size class are many of Volvo's traditional suppliers, located in the small firm 'Gnosjö' district southwest of Göteborg (see Chapter 12).

Table 6.2   Numbers of domestic sub-contractors to the Volvo Torslandaverken assembly plant (Göteborg), by product category and firm size, 1997

	Size group (employees)						
	1-19	20-49	50-99	100-249	250-499	500	Total
Metals	8	11	6	13	7	5	50
Plastics/rubber	3	2	3	11	6	1	26
Chemical/paint	5	4	4	2	1	0	16
System/module	0	1	2	2	1	3	9
Electronics	1	2	1	0	1	1	6
Textiles	1	0	1	2	0	0	4
Retail	0	2	0	0	1	0	3
Service	2	0	0	0	0	1	3
Other	1	1	4	0	4	1	11
Total	21	23	21	30	21	12	128*

* The difference in total number of sub-contractors between Tables 6.1 and 6.2 refers to 26 companies it was not possible to classify by product and/or size due to lack of data.

Source: Data compiled by the authors

Among the larger sub-contractors there is no direct geographical pattern pointing towards agglomeration or dependence on Volvo and the Göteborg facility. Compared to the study by Törnqvist, this spatial pattern is much the same as in the early 1960s. The most significant change is the large reduction in number of suppliers. The concentration of suppliers in the southern part of the country reflects Volvo's increasing demand for just-in-time deliveries to meet its own time schedule.

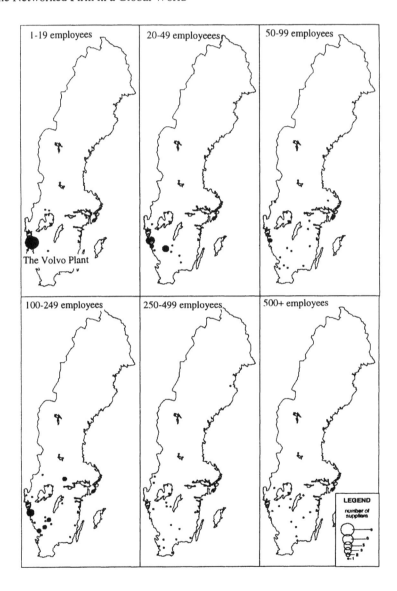

Source: Data compiled by the authors

Figure 6.2   Regional distribution of sub-contractors to Volvo's Torslanda plant by size group of suppliers, 1997

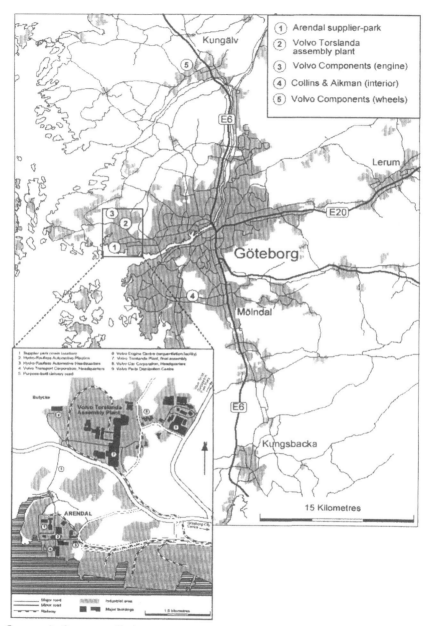

Source: Authors compilation

Figure 6.3 Location of sequence suppliers to Volvo's Torslanda plant in the Göteborg region and inside the Arendal supplier park

The case of Volvo Arendal supplier park

Background

To meet the increasing time related demands on automotive buyer-supplier relations there have been several attempts to secure reliability through the development of 'supplier parks' adjacent to main assembly plant. One model has been the agglomeration of suppliers in Toyota City, a textbook example of how spatial proximity has been used to ensure control over the production process. European examples are more scarce and small-scale. Fiat's new plant in southern Italy (Pulignano 1997) and Ford's plant in Valencia have core suppliers close to those assembly plants (Financial Times 1998). Another example is the SMART-project, including Daimler-Benz and the Swiss watch-making company SMH, to build a small car in a completely new factory with the major suppliers located within the same building complex (Financial Times 1997).

An example of a greenfield automotive plant with a supplier park strategy is the Ford/VW project south of Lisbon in Portugal (Ferrão and Vale 1995). The location, termed 'Fordland', had 20 first tier suppliers in 1993, created by the demand for JIT deliveries. However, the suppliers in 'Fordland' had little impact on the local economy. They were either linked to an international company or functioned as a delivery warehouse for traditional suppliers located elsewhere.

In Sweden, a supplier park has been in operation since 1994 associated with Saab's automotive assembly plant in Trollhättan, 70 kilometres north of the Volvo Torslanda plant. Saab is at present 50 percent owned by General Motors and has in recent years become closely integrated within the GM European production system. The supplier park project was initiated by an independent business consultancy. Its intention was to attract and provide business services to suppliers, mainly linked to Saab. The project led to the creation of a new firm, Se-De International, with responsibility for the co-ordination of sequential deliveries to the Saab factory. From the initial aim to attract automotive suppliers to the region, there has been a change of direction in the work of the company. Today they act as a third-party logistics firm with their main business in the co-ordination and sequencing of the flow of components from approximately 20 European suppliers to Saab. Se-De International had 50 employees and an estimated turnover of 31 million Swedish Kronor (£2.5 million) in 1998.

The benefits from this arrangement were twofold; flexibility in supplier selection, and the possibility for the logistics firm to co-ordinate transport more efficiently. A number of the suppliers have extended the use of the sequencing warehouse and moved some of their final assembly there. One example is the automatic welding of exhaust systems by a German supplier who previously shipped the finished systems from Germany to Trollhättan. Today, it is possible to ship the parts of systems much more efficiently and thereby minimise transport costs, delivery issues and quality problems.

In contrast to Saab, Volvo itself initiated the supplier park. Its purpose was to attract physical production to the area. Transport and other services can be provided by VTC, which is located within the same area. Volvo owns the land on which the supplier park is located, a fact that probably simplified the planning and co-ordination of the process of attracting suppliers.

Because its production is located in Sweden and Belgium, Volvo needs a highly efficient and reliable transport system on a European scale. The Göteborg assembly plant receives more that 50 percent of its input components from outside Sweden, which always includes sea transport in some form. To guarantee reliability, Volvo uses a 'milk-round' system for the pick up of goods from suppliers in Sweden and Europe. This arrangement is supposed to guarantee a 48-hour transport time from any European supplier to the Volvo assembly plants in Göteborg and Ghent.

This means that suppliers have one week to perform the following sequence:

- receive and process the final order;
- procure inputs and manufacture the goods; and
- pack and prepare for transport to Volvo.

On these grounds alone, there are obvious advantages in being located close to the vehicle assembler. The shorter the time for transport, the longer time the supplier can use for the other parts of the process without having to rely on large stocks. Most European suppliers can meet this demand through Volvo's pan-European transport system.

In addition to these time and delivery issues, production is also being modularised. Volvo wants to assemble modules developed and produced by a number of large suppliers located close to the Volvo plant. The supplier takes over assembly work and responsibility for product development that was formerly done in-house by Volvo, and delivers a more complex and high value product ready to be assembled into a car.

Because these modules are high value, voluminous and have many variants, storage is expensive in terms of space and the capital tied up in finished products. Instead, supply is organised through sequential synchronised deliveries with minimum inventory. The result of this arrangement is that suppliers of modules have to deliver with extremely high frequency and in small lots. The temporal constraints in this type of arrangement are very tight and spatial proximity is one of the strategies that Volvo is using to secure reliability. The main incentives behind the supplier park project are thus:

• a general objective to achieve a three-week lead-time from customer order to delivery across the entire European production system. This involves a pan-European transport system within which a supplier park functions as a hub;
• the trend towards sequential delivery and modularization. A selection of suppliers will take on more responsibility and have to deliver according to sequential JIT requirements.

The development process of the supplier park project

The first steps towards establishing a supplier park were taken in the early 1990s by outside actors, mainly construction companies, who used proximity to the Volvo plant as a way of developing and marketing land as industrial estates. They were actively promoting the sites as a 'just-in-time' supplier park. However, the results of these efforts were relatively poor, and did not attract enough interest from suppliers to warrant the establishment of a site dedicated to the automotive industry (Larsson 1993).

Volvo itself expanded its logistics activities during the late 1980s and started to develop a transport terminal at the former shipyard site of Arendal, located just five minutes by truck from the Torslanda plant. This area provided enough space and was close enough to be developed as a supplier park.

The process of creating a supplier park started within Volvo in 1996 as a part of the development of the new S80 model. At the time, the Arendal site was occupied only by VTC and Hydro-Raufoss, who used one of the buildings for storage and sequencing the delivery of bumpers, transported from Norway and Belgium.

During 1997, all the industrial buildings on the site were upgraded and transformed to suit the needs of Volvo's preferred suppliers. This was done in parallel to selecting and attracting the suppliers of components, which needed to be delivered in sequence to the new S80 model. In the second half of 1997, all suppliers were located on the site, and Volvo had developed 35,000 square metres of new industrial space.

Being located in the park was clearly a condition for the award of a contract by Volvo. Few suppliers deliver in sequence from other locations (see Figure 6.2). Those that do had previously delivered components in sequence to the S/V 70 models. Production of the S80 model started in April 1998, and from this time suppliers were expected to deliver according to Volvo's JIT requirements.

Function of the supplier park

Firm size and products

At the time of the production start-up of the new S80 model, there were eight suppliers located in the park. The main aim of these units was to prepare for sequential deliveries by sorting, and in some cases pre-assembling incoming components. In Table 6.3, the name, country of origin, product and number of employees of the units located on the Arendal site are listed.

In terms of size, the firms can be divided into two groups, Hydro-Raufoss and Lear with 100 or more employees and the other six, which are considerably smaller. This is explained by the fact that the operations of the two larger companies are significantly different from the others. Hydro-Raufoss runs a complete production process, and Lear Corporation used, until the abandonment of its Bengtsfors plant in early 1999, the Arendal site to store and sort components not delivered in sequence. Among the six other suppliers, Delphi-Packard is developing a multi-product facility, while the rest are dedicated one-product companies.

The products sequentially delivered from the Arendal site are typically:

- voluminous, involving large storage spaces and high unit transport costs;
- model-specific and/or colour-specific components, with large numbers of variants that are difficult to pre-produce and store in terms of both space and fixed capital.

The majority of products fall into the second category and are related to the interior design or the colour of the car (for example, painted bumpers). 'Non-visible' components, such as exhaust systems and fuel tanks, fall into the first category.

Table 6.3   Production units by home country, product and number of employees located at the Arendal supplier park, June 1998

Company	Country	Product in sequence	Employees
Hydro-Raufoss Aut.Plastic	Norway	Bumpers	120
Lear Sweden Interior Syst.	US	Door panel, dashboard	100
Delphi-Packard	US	Cables, electronics	50
Becker Sweden AB	US/Germany	Interior roof, tunnel console	40
Walbro Automotive	US	Fuel tanks	25
Borgers Nord AB	Germany	Trunk carpet, parcel shelf	20
Tenneco Automotive	US	Exhaust system	20
Rieter Automotive	Switzerland	Carpets	10

Source: Data compiled by the authors

In the current debate on the automotive industry, much attention has been paid to the development to out-sourcing and the transfer of responsibility to suppliers. This is clearly reflected in the new supplier park concept. Approximately 60 percent of the value of Volvo's purchases come from the eight Arendal suppliers plus Volvo Components (engines) and Lear Seating. This indicates the high value of the products. Volvo calls them module suppliers, reflecting the complexity of the product involved. An example of such a product is a front bumper module, which contains the following components: aluminium structure, plastic foam, painted plastic shield (in the colour of the car), spoiler, lights and sensors. Just two car-generations ago, all these items were put together by Volvo on the final assembly line. All sub-contractors at the Arendal sites are module-suppliers, although with different levels of complexity and value added.

Production process and technological level

The module suppliers on the Arendal site are delivery facilities rather than production facilities owing to the complexity of the components they deliver. The only facility with on-site production is Hydro-Raufoss, who recently upgraded their operations from a warehouse to a complete production unit. Among the other suppliers, there is a range of activities from pre-assembly of components (such as welding of exhaust-systems) to simple storage and sequencing of finished products.

Table 6.4 shows the type of activity and the geographical distribution of purchases for the production units located in the supplier park. It is evident that the majority of the suppliers have a relatively simple production process, which includes only marginal value adding to the incoming products before they are sorted and delivered in sequence to final assembly at Volvo.

There is one company which does only sorting and sequencing without any assembly at all, while the majority of suppliers perform pre-assembly to some extent. This can range from relatively simple adding of clips and fasteners, to more complex automatic welding of exhaust pipes or assembly of displays into a complete dashboard.

Table 6.4   Type of activity and relative distribution of purchases (value) for production units at the Arendal supplier park, June 1998

Company	Type of activity	Product develpm.	% input local	% input Sweden	% input Europe	% input world
Hydro-Raufoss	production	yes	0	3	95	2
Becker	pre-assembly	yes	2	10	88	0
Borgers	pre-assembly	yes	5	30	65	0
Rieter	sequentiation	no	4	0	96	0
Tenneco	pre-assembly	no	0	35	65	0
Lear	pre-assembly	no	5	80	15	0
Walbro	pre-assembly	no	0	0	85	15
Delphi-Packard	pre-assembly	yes	0	0	100	0

Source: Data compiled by the authors

As indicated in Table 6.4, four of the suppliers have product development located at the Arendal site. The extent and content of these activities has not been studied, but there are only limited Volvo-specific product development functions in the park. The only major exception is Hydro-Raufoss that moved their entire development function to the Arendal site in 1997 in connection with their investment in a new production facility. Lear Corporation has their product development function located elsewhere in Göteborg, together with a number of production facilities across southern Sweden, acquired during the late 1990s.

Purchasing pattern and local linkages

Table 6.4 also shows Volvo's low level of local and national sourcing. The Arendal facilities only sort and deliver ready-made products usually made elsewhere in Europe, except in the case of Lear. Lear has a national presence in Sweden. Some of the other cases of national purchasing are explained by internal transactions from other units located in Sweden. Some suppliers have a long tradition of supplying Volvo and have built up a relatively large supply base in Sweden.

A significant aspect of Volvo's purchasing is the extremely low level of local sourcing. Local in this context refers to Göteborg and adjoining local administrative districts (see Figure 6.3). Indeed, no major local supplier has been involved in establishing the supplier park. Volvo has, in fact, commented on the "very low level of interest from local domestic companies to establish contacts with the new module suppliers" (pers. comm.) It has to be noted though, that this process is in its initial phase, and local linkages may develop to a greater extent when the situation has consolidated.

The domination of European suppliers reflects the production structure of the companies in the park. Components are produced at their existing locations in Europe as before, and shipped in batches to Göteborg. However, these arrangements may represent only the start-up phase of module supply, and local links may develop in the future.

Organisation of transport

The single most important aspect of the Arendal supplier park is the organisation of high frequency, sequential deliveries from the suppliers to the Volvo assembly plant. This involves extremely high levels of reliability in the transport function and is, therefore, the responsibility of a wholly Volvo-owned company, VTC. The frequency of delivery for the sequenced components is normally 6-10 times per day. This means delivering every hour during full-scale production. Some suppliers, such as Lear who delivers non-sequence modules, do have significantly more deliveries per day, although the time-restrictions and numbers of variants are not as rigid as for the sequenced deliveries.

As already discussed, the main function of the supplier park in this initial stage is to handle the logistics of incoming batches of component parts from the European continent, and to transform these shipments into car-specific components to be delivered hourly. These logistics and transport functions were one of the areas where Volvo expected to gain scale economies by offering to handle all suppliers' incoming transport. The situation today, however, includes a mix of arrangements. For example:

- the local unit is responsible for logistics organisation and uses outside firms or Volvo for physical transport;

- the parent company in Europe is responsible and has a contract with VTC for the organisation and physical transport of components; and
- the second-tier supplier is responsible for deliveries to the first tier supplier at the Arendal site.

Independent of the form of logistics organisation, there is a need for co-operation within the park in order to utilise the space as efficiently as possible and to minimise the negative environmental impact of the generated traffic. According to Volvo, this is an issue where internal co-operation between suppliers is needed. There are potential economic and environmental benefits to be gained if companies co-ordinate their shipments of incoming materials from the European continent.

Power relations

An interesting aspect of the restructuring process in the automotive components sector is the concentration of ownership on both a global and national scale. Small local units now belong to large international or global corporations, which significantly alters the power of a relatively small automotive manufacturer such as Volvo.

From Table 6.5 we can see that total sales for the companies behind the eight Arendal suppliers amounted to US$ 58 billion in 1997, which is more than twice the turnover for the entire Volvo Group. The single most important parent company is Delphi Automotive, followed by the Norwegian petrochemical and metals group Norsk Hydro, who owns Hydro-Raufoss Automotive as part of its light metals division. This situation reflects the combined processes of concentration and internationalisation in the automotive supplier business. This restructuring process has affected Sweden relatively late. A study by the Swedish Industry Board in 1990 (SIND 1990) presented 15 large groups as important owners of automotive suppliers in Sweden. Out of these only four were foreign-owned. This picture has changed dramatically over the last few years, as foreign supplier multinationals have acquired many of the formerly domestically owned first tier suppliers.

For the S70/V70 model, which is the generation preceding the current S80 model, there were only two major Swedish-owned supplier groupings among the 40 most important domestic suppliers. Global actors, such as Lear Corporation and Collins & Aikman controlled ten of the former Swedish-owned larger suppliers. This is even more pronounced in the development of the supplier park as indicated by Table 6.3. New global firms, such as Walbro, Delphi and Tenneco, have entered the list of suppliers with the introduction of the S80 model. Together with the other firms in the park, they supply almost half of the value of the car. So far, the take-over of the domestic parts industry by multinationals has, with the exception of the move of Lear Seating, not resulted in any spectacular changes at the plant level. On the contrary, the take-overs may in the long-term create opportunities to supply larger markets from the Swedish base, given their ability to specialise and to adapt to the requirements of their new majority owners.

Table 6.5   **Total sales and numbers of employees in 1997 for the mother companies of suppliers to the Volvo Group**

Local sub-contractor	Parent company	Sales ($US m)	Employees
Delphi-Packard	Delphi Automotive Systems	28 400	216 000
Hydro-Raufoss Auto.Plastics	Norsk Hydro ASA	12 000	38 000
Lear Sweden Interior Syst.	Lear Corporation	7 342	50 000
Tenneco Automotive Swed.	Tenneco	7 200	50 000
Rieter Automotive Sweden	Rieter Group**	1 320	12 300
Becker Sweden AB	Becker Group*	1 300	8 400
Walbro Automotive	Walbro Corporation	620	4 700
Borgers Nord AB	Borgers GmbH	323	2 700
Total		58 505	382 070
Volvo Group		22 953	73 000

* Becker Group US and Becker Group Europe were acquired by Johnson Controls in April 1998. Figures for Johnson Controls are not presented in the table. Sales figures are estimates for 1998.
** Figures for Rieter Group are for 1996.

Source: Official company internet sites, PR Newswire-company news Internet service

The out-sourcing of more production and development functions to larger global suppliers may also pose problems for the assembly firms. The problem for the principal assembler could, in the long run, be that suppliers develop competencies and financial resources that put them in an extremely favourable situation in negotiations over future projects.

Another consequence of Volvo's suppliers being owned by large, foreign mutlinationals is the possibility of closure when there are problems or when a project or contract is finished. The investments in the Arendal-site are marginal relative to the total investments of these companies, with the exception of the Hydro-Raufoss investment in a new production plant. Given that only 600 jobs will be created in the supplier park up to 2001, the rapid process of setting up operations could easily, given the types of activity involved, be reversed at the same speed if conditions change.

The supplier park concept and the development of a creative environment for automotive-suppliers

Introduction

A recent debate in economic geography focuses on the importance of industrial agglomeration for regional development. In a review of the research within this field, Malmberg et al (1996) make a distinction between forms and forces behind industrial agglomerations.

We can summarise the forms of agglomeration as being either of a general type or consisting of related firms and industries. This distinction is important because it indicates the significance for growth of the internal division of labour and the degree of functional linkages within a region. The forces behind the creation of such agglomerations have traditionally been explained in terms of transaction cost efficiency and system flexibility. This approach has been challenged by the introduction of knowledge, as being just as important as costs in today's highly competitive industries.

The basic line of argument put forward by Malmberg et al (1996) is that the regional clustering of related industries and firms facilitates knowledge accumulation. This is especially important in relation to the types of information that can not be standardised and communicated over distance, where factors such as trust and common social values are important.

In the case of the Arendal supplier park, this framework can be used to discuss the extent to which it is possible and/or desirable to promote the development of a 'creative environment' in such a location, in the sense of knowledge distribution and generation within the automotive components industry. The current situation, though, differs significantly from the theoretical prerequisites discussed in the literature. First, the rationale behind the agglomeration of suppliers is the demand from Volvo to locate close to the assembly operation. This has resulted in, at least in the initial phase, a situation where all contacts are made through Volvo. There is no internal co-operation or division of labour among supplier park plants. The only areas where transaction costs might be saved are on transport arrangements. Second, linked to this point, is the lack of co-operation between related firms. Because all the suppliers perform the same function, there is no incentive for joint development projects and knowledge accumulation within the supplier park.

With these remarks in mind, it is obvious that the driving force in this particular case is the strategy formed by Volvo, which currently is focused on the day-to-day problems of production co-ordination and physical delivery. To move from the current situation towards a strategy with more internal co-operation and joint development projects, creating an environment for knowledge accumulation, future competitiveness and regional development, there are some basic requirements. First, there is a need for related and supporting industries to locate in the area in order to broaden the base for knowledge accumulation. Second, more development, and other knowledge-intensive functions need to be present. A prerequisite is that local social networks and local institutional conditions are also favourable for co-operation and communication - including institutions, such as local authorities and institutions of higher technical and business education.

Related and supporting industries

As might be expected from Volvo's strategy, there are very few firms in related businesses such as sub-contractors or service providers located in the Arendal park. There is only a handful of companies providing services directly to suppliers. One of these companies is an engineering consultant specialising in the automotive sector, while the others deliver relatively uncomplicated services, such as security or office supplies.

The most important actor is VTC with its headquarters and transport terminal located in the supplier park area. This company's activities are centred on the co-ordination of Volvo shipments to and from Sweden. With respect to the suppliers located in the park, VTC is responsible for all the sequential deliveries to the Torslanda plant. Some of the suppliers also use VTC for their in-bound deliveries. There is also a service for smaller suppliers who deliver in sequence from other locations, whereby VTC holds a stock of components for sorting and delivery. Transport and logistics might be an area where the park can in future attract and develop new businesses thanks to its proximity both to other suppliers and also to VTC.

Moving beyond the individual supplier park to the regional scale of southwest Sweden, there is however a clear regional concentration of assembly plants and suppliers in the automotive industry. Volvo, as well as Saab, have their domestic assembly plants in the region and are deeply involved in the regional economy.

Product development and knowledge accumulation

A prerequisite for the transformation of the Arendal supplier park from a logistics centre into an environment for knowledge accumulation is the local presence of knowledge-based activities. Volvo certainly has a long record of co-operation with local institutions of higher education, especially Chalmers University of Technology, which educate many of the engineers employed by the company. The creation of such an environment might also come about if existing companies and independent firms locate more product development in the park.

At present, however, only half the park's firms (Table 6.2) report that they undertake product development on-site. The others have at least one person posted inside Volvo working on development projects. But, this is development concerned principally with unique Volvo products and the benefits from co-operating within the supplier park are limited. So, the situation in Arendal today is one of firm-specific knowledge accumulation, where each unit in the supplier park either has its own responsibility or has to work through its foreign mother company. This is not an environment that leads to local knowledge accumulation.

Conclusions

The objective of this chapter has been to examine the development of the Arendal supplier park project, with special emphasis on the relationship between production strategy and spatial proximity, and focusing also on the issue of firm size. Furthermore, we have discussed the supplier park idea in terms of the possibility of developing a creative environment for knowledge accumulation within the automotive supply industry.

Volvo's co-ordination of its production process has been the main driving force behind the Arendal supplier park project. It has resulted in the establishment in the park of module suppliers with very limited production functions. Their roles are, instead, to store, pre-assemble, sort and deliver modular components, in sequence, to the Volvo assembly plant located three kilometres away. Thus, spatial proximity is a basic rationale behind the location of the supplier park to meet Volvo's requirement for high frequency and reliable deliveries.

All suppliers are small or medium sized in terms of on-site employment, but they are all parts of foreign-owned supplier multinationals. Processes of international concentration have created fewer and bigger supplier groups in the automotive sector, significantly changing power relations within it. The Swedish motor vehicle industry has been affected by these processes relatively late, and has only recently come to adjust to domestic supplier groups having been taken over by international companies.

From the perspective of the principal firm, in this case Volvo, there are clear tendencies towards the spatial concentration of its first tier suppliers in close proximity to its assembly plant. At the same time, the specialist sub-contractors have internationalised and have become integrated into highly dispersed networks of transnational automotive suppliers, though operationally they are split up into small and medium sized branch plants, tailored to meet the demands and requirements of final assembly firms.

The potential to develop a creative environment in the supplier park is obviously limited. Most important is the fact that Volvo drives the entire project with the specific purpose of co-ordinating the physical deliveries of components. The issue of knowledge accumulation, inter-firm co-operation and regional development has not been on the agenda. This also means that the companies in the park all have the same function, to serve Volvo, and have no direct incentive to co-operate with each other. Predictions on what the environment of the supplier park will be like in the future, however, are difficult to make.

What is clear, however, is that physical distribution and production still matter in geography. The case study presented in this chapter demonstrates the impact of information and communications technology and other modern tools of communication as a complement to, rather than a driving force of, industrial location. Organisational as well as physical forces are fundamental to understanding why production is located where it is. What in this Swedish case study might initially appear to be small and medium sized suppliers are, in reality, branch plants of giant, global operators. Ultimately, the power relations between Volvo and it's first tier suppliers located at Arendal, might not be to Volvo's advantage.

7 The Unresolved Question of New versus Old: Technological Change and Organisational Response in the German Chemical Industry

Harald Bathelt

Introduction[36]

Since the late 1980s, economic power has become progressively concentrated within a number of industries at an international level. In the chemical industry, a number of mergers and acquisitions have, for instance, been undertaken by large multinational firms to establish even larger conglomerates with production and marketing facilities in each of the major world regions. The question of how and the extent to which economic concentration and globalisation tendencies will influence the production programs and processes and the supplier and customer relations of individual chemical firms has yet to be answered. It is also unclear how mergers and acquisitions of large firms affect the market position of small and medium sized firms in the chemical industry. Will SMEs for instance, be increasingly integrated into the hierarchical production system of large producers? Or, will they be able to approach and occupy new market segments as large firms concentrate their activities in a limited number of core areas? To what degree will SMEs be forced to redefine their market strategies and restructure production in order to strengthen their competitiveness? Will new opportunities, as discussed in the preceding chapters, be created for such firms as their larger counterparts reduce their presence in traditional market regions in favour of global production systems?

This chapter will address these closely interrelated questions. Size alone would not serve as an adequate variable to differentiate firm strategies and behaviour beyond describing a superficial small versus large firm dichotomy. A more completed typology of firms will therefore be developed. This has also been emphasised in chapter two. Already Taylor and Thrift (1983a) have identified segmentation patterns in business organisation that go well beyond a simple dualistic structure. In chapter four we states that different types of sub-contractors have different opportunities in the internationalisation process. As the chemical industry is an extremely heterogeneous industry group, the analysis presented here will be based on one particular branch of this industry, that producing pigments, dyes, paints and varnishes (PDPVs).

Structural change, localised learning and the division of labour[37]

To better understand the underlying processes, the above-mentioned issues should be viewed within the context of the structural crisis that has affected the economic and societal structures in Western Europe and North America since the 1970s and 1980s. From a regulationist perspective a crisis in Fordist development has dominated in the post-World War II period (Lipietz 1987, Boyer 1990). The economic and societal problems and contradictions of the Fordist crisis, such as mass unemployment and deindustrialisation tendencies, seem to have particularly affected certain places and regions (i.e. the locations of large firms in mass-producing industries and their suppliers). This indicates that the Fordist crisis has a geographical bias (Moulaert and Swyngedouw 1989). Firms that are located outside the traditional manufacturing areas in new production spaces are seemingly better able to cope with changes in economic, technological, institutional and societal settings (Scott 1988b).

According to regulation theorists, a recovery from the Fordist crisis will require a new regime of accumulation and/or mode of regulation to ensure coherence and reproductivity of the economic and societal structure. The new regime must have the capacity to overcome the technological, economic and social limits of Fordism to be successful (Jessop 1992). A post-Fordist mode of development is often assumed to be characterised by flexible technologies and production processes, flexible labour organisation, segmented markets, a volatile demand structure and decentralised co-ordination patterns. It is interesting to note that similar conclusions have been drawn by neo-Schumpeterians. They view the present crisis as an indication of a shift in the techno-economic paradigm (Freeman and Perez 1988). Scholars from both schools of thought believe that a recovery from the Fordist crisis will require that flexible ones replace rigid Fordist practices and structures in the industrial sector.

In line with the flexibility debate, various scenarios have been developed to describe the potential future of the technical, social and spatial division of labour. These scenarios assume combinations of different forms of flexible technologies, labour and production processes to yield models of future production. The most influential scenarios are those of flexible specialisation by Piore and Sabel (1984) and dynamic flexibility by Coriat (1992). Both scenarios are, however, of limited applicability for they describe idealistic forms of flexible production, which depend on specific market structures (Bathelt 1995b). Along with Piore and Sabel's (1984) work, many studies have postulated a growing importance of networks of small and medium sized firms in overcoming the Fordist crisis as discussed in chapter two. Such statements have ignored the fact though that a widely accepted definition of what a small firm is does not even exist.

One particular problem of previous empirical studies about flexibility processes has been the virtual lack of an attempt to distinguish between the temporary strategies of firms and permanent structural changes (Jessop 1992). It is, therefore, still largely unclear which product, process, linkage and spatial structure may lead to a flexible post-Fordist (or a not yet pre-determined after-Fordist) mode of development. The degree to which the new development mode will depend on flexible processes is also open. Scholars, such as Amin and Robins (1990), Gertler (1992) and Sayer and Walker (1992), also criticise the selective and contradictory nature of the flexibility literature and question the significant of flexibility processes. In recent study, Harrison (1997) provides detailed evidence that large firms continue to dominate global economy. He identifies tendencies toward 'concentration without centralisation' as being important within many industries. Harrison (1997) believes that the popular emphasis on the success of SMEs growth and their ability to respond to the changing settings of the Fordist crisis have been overemphasised.

In his investigation of international trade flows, Storper (1992) found that many countries are specialised in the export of particular commodity groups and their production is spatially concentrated in certain regions. According to Storper (1992, 1997), firms within these commodity chains obtain a specialised technological competence from continuous learning processes. Product-based technological learning strengthens the firms' competitiveness and provides a basis for export activities. This process is further enhanced by the development of territorial production systems (technology districts) which are characterised by networks of small and medium sized, vertically disintegrated firms (or by more or less hierarchical linkage systems of large, vertically integrated firms). Within these technology districts, collective learning, flexibility and proximity are closely interrelated and stimulate one another (see also Morgan 1997). Cumulative technological learning and intra-regional production linkages are based on relatively homogeneous goals and attitudes and on generally accepted rules and conventions with respect to the technologies used, the resources to be mobilised, inter-firm co-operation and capital-labour relations. From this, product and process based competencies can develop within the technology districts and serve as a basis for trade specialisation (Storper 1992, 1997).

Maskell and Malmberg (1997) have also stressed the importance of learning processes in a local context. They argue that a firm's competitiveness depends on a unique set of competencies and its ability to develop them further through continuous learning processes. If such firm-specific competencies are based on localised capabilities (such as specialised resources and skills and shared trust, norms, routines, traditions and other local institutional structures), a regional competitive advantage will result. Due to the evolutionary character of knowledge generation, firms and regions with the most sophisticated skill levels, know-how and research activities have the best opportunities for further knowledge creation. This attracts specialised economic activities to a region and stimulates cumulative regional growth. As a consequence, this may lead to regional specialisation and concentration that would, in turn, strengthen existing localised capabilities.[38]

Observations of international trade patterns, similar to those of Storper (1992), have led Krugman (1991) to view the regional (subnational) level as a catalyst for industrial concentration and specialisation. In contrast to Storper (1992), he emphasises economies of scale and, thus, highlights the role of large firms. He views regional economic development as a historical process through which locational patterns are continuously produced and reproduced. In his explanation of local specialisation tendencies, Krugman (1991) draws heavily on Marshall's (1920) classic economic analysis of industry localisation. In that analysis, local specialisation is a product of three factors: (1) labour markets pooling (the existence of a large pool of and demand for specialised labour), (2) advantages due to the agglomeration of specialised suppliers (associated with increasing returns and low transportation costs), (3) technological spillover effects (synergies from a particular 'industrial atmosphere' which stimulates the creation and dissemination of knowledge).

Krugman (1991) views technological spillover as being the least important of the three factors as it tends to be limited to high technology industries. In this respect, his evaluation differs substantially from Scott's (1988b) analysis of agglomeration tendencies. Krugman's (1991) emphasis of economies of scale and transportation costs might be useful in the explanation of the importance of distance in the supplier and customer linkages in some branches of the chemical industry (such as basic chemicals). It fails however, to provide a deeper understanding of the cohesive forces in other branches (such as PDPV). Evidence will be provided here that technological spillover, inter-firm communication and interactive learning processes are not restricted to high technology industries.

Within this context of industrial restructuring and localised learning, empirical evidence will be presented in this study on the direction of changes in the product and process structures of PDPV firms, on their ability to utilise supplier and customer relations to increase their competitiveness and on the role of proximity in these changes. Based on a postal survey and in-depth case studies of the German PDPV industry, the question of how firms have adjusted their production programme and processes to meet the changing economic, technological institutional and societal settings of the Fordist crisis will be addressed. The analysis involves a combination of both quantitative data from a postal survey and qualitative data obtained from intensive interviews with selected firms. The questionnaire data was used to establish a firm typology according to various product and process characteristics. Special emphasis was given to the exploration of technological change, flexibility processes and producers' buyer-supplier relations and their spatial implications.

The German PDPV industry

The empirical work presented here is part of a larger study of the German chemical industry (Bathelt 1997).[39] Despite its importance within the manufacturing sector, the chemical industry has been largely neglected in analysis about the changing nature of industrial production in response to the Fordist crisis.[40] The sociological studies of Kern and Schumann (1990) and Schumann et al (1994) about workplace, technology and labour organisation in the German chemical industry are an exception.

The PDPV industry is focused on in this chapter for several reasons. First, the production of pigments, dyes, paints and varnishes has been historically central to the chemical industry and has provided a basis from which other product groups have evolved. Second, the PDPV industry produces both intermediate and final products. Unlike most other chemical branches, it is, therefore, also oriented towards end users and consumer markets. Third, the PDPV industry has remained a stable industry group with above average growth rates and a high degree of international competitiveness. Fourth, the technological configurations in the PDPV industry are not as rigid and complex as they are in many other chemical branches. Product and process structures can be adjusted more easily to meet changes in technological, economic, societal and institutional settings. Therefore, the PDPV industry has greater potential for product and process flexibility in comparison to the other branches investigated. In showing that flexibility processes are relatively unimportant in this industry group, it will be argued that flexibility is of minor significance in successfully overcoming the Fordist crisis.

The spatial structure of the German PDPV industry is characterised by a distinct pattern of concentration. Four primary locational centres of the PDPV industry can be identified: Cologne-Dusseldorf-Wuppertal, Frankfurt-Wiesbaden, the Neckar Valley and the Hamburg Region. These centres correspond to the locations of some of the largest German chemical firms such as BASF (Ludwigshafen), Bayer (Leverkusen), Henkel (Dusseldorf) and Hoechst (Frankfurt). This spatial pattern is closely related to the historical development of the chemical industry, especially its rapid growth in the last century (Henneking 1994, Muller-Furstenberger 1995, Bathelt 1997). At that time, massive population increases stimulated the demand for bleaching-agents and dyestuffs. Around 1850, a large number of dyestuff and paint procedures were established to satisfy these needs. Many of the original location decisions were made within a relatively wide range of spatial opportunities rather than being predetermined by favourable location factors (Schall 1959). The early start-ups played a decisive role in the development of the chemical industry. They determined the growth path of the industry because of their strong R&D tradition in core technologies and their high level of technological expertise in product and process innovations. These firms took advantage of economies of scale and technological synergy. They grew into large, extremely heterogeneous, vertically and horizontally integrated organisations. The dominant producers, especially BASF, Bayer and Hoechst, became leaders in practically all branches of the chemical industry (Schall 1959, Freeman 1990, Bathelt 1995a).[41]

In terms of its size, the PDPV industry can be classified as a relatively small branch within the German manufacturing sector. In 1994/95, the industry consisted of 380 predominantly small and medium sized firms. The PDPV industry had a total labour force of 52,000 in 1996 (Statistisches Bundesamt 1997, VLI 1997). Of these, 23,000 people were employed in producing pigments and dyes, with the remainder in other coatings and related products. Employment in the industry has remained relatively stable during the 1990s (VLI 1994, 1997), at a time when other sectors have suffered from substantial job losses (VCI 1994b, 1997).

In 1995, German exports and imports of paints and varnishes accounted for 277,000 and 123,000 tonnes respectively (VLI 1997). Despite the fact that export volumes were more than twice as high as imports, the latter had much higher growth rates during the 1980s and 1990s (VLI 1988, 1994). This was not, however due to an increase in international price competition. This was exemplified by the fact that 98 per cent of all German imports were acquired from within the EU (especially Great Britain, France and the Netherlands), as opposed to low-cost countries in Eastern Europe and East and Southeast Asia. Overall, German PDPV producers have been successful in maintaining their position as world leaders in many product groups (Bathelt 1997). Between 1984 and 1992, producers of paints and varnishes in Western Germany increased their share of the value of total chemical production from 7 per cent to 9 percent (VCI 1993, 1994b).

Since the 1970s, a number of changes have taken place in the economic, technological, institutional and societal settings of the PDPV industry. These have stimulated firms to rethink and reorganise their production programmes and market strategies. The changes have also given rise to the development of new products and markets. Producers of paints and varnishes for the construction industry have, for instance, experienced a high growth rate in sales since the late 1980s. This is due to German unification and especially to the growing demand in Eastern European countries. In other market segments, such as textile dyes, German producers have had to cope with a decrease in demand due to the tendency of textile firms to relocate to East and Southeast Asia. Further shifts in demand towards East and Southeast Asian markets are expected in the future (Howard, 1996). In the automobile industry, producers have introduced new production technologies, along with new production concepts such as total quality management (Bertram and Schamp 1989, Hanack 1995). This has resulted in a reduction of the amount of paints and varnishes applied per car and an increase in competition between the PDPV producers.

To cope with the changes in demand patterns, the PDPV producers have sometimes incorporated new process technologies (e.g. *Chemie-Produktion* 1997). These have increased the quality of outputs, reduced the amount of raw materials needed and allowed for greater flexibility in production. The latter is especially important since individualisation tendencies in consumption have resulted in further market differentiation and fragmentation. In addition, environmental regulations have tightened and consumers have developed a distinct preference for 'environmentally friendly' products. This has put pressure on PDPV producers to reduce the amount of toxic solvents and heavy metals in their products (Howard 1996).

Methodology

The empirical part of the present study has followed four stages: the establishment of a data base, the execution of a postal survey, the identification of organisational clusters, and the completion of case study analyses. In the first stage, various manufacturing directories (Hoppenstedt 1993a, b, c, d, VCI 1994a) were analysed to establish a database for PDPV producers in Germany. Overall, 360 and 20 PDPV firms were identified in Western and Eastern Germany respectively. Of the total of 1,870 German Chemical firms identified, about 20 per cent had production facilities in the PDPV industry (Bathelt 1997). Firms in Eastern Germany were excluded from further analysis to avoid the potential for changes in the economic and societal system to introduce bias into the analysis.

In the second stage, questions related to how PDPV firms have adjusted their products and processes to deal with changes in the technological, economic, societal and institutional settings were investigated. In this stage, a standardised questionnaire was developed and a postal survey conducted between September 1994 and February 1995. Nearly half of the firms previously identified (169 of 360 producers) were asked to participate in the postal survey. A total of 60 firms participated in the survey. This corresponds to a response rate of 35.5 per cent, which is relatively high. The results of chi square tests did not indicate significant differences in the size, age and spatial distributions between the total and sample populations (Bathelt 1997).

The objective of the third stage was to derive a typology of PDPV firms, which would encompass the major configuration and tendencies in the technical division of labour. This was done using multivariate classification techniques. The survey data was transformed into a set of binary variables. Various cluster and discriminant analyses were then performed to identify relatively homogeneous groups of firms in terms of their product and process organisation. These groups will be referred to as organisational clusters in the remainder of the chapter. Before undertaking the classification process, variables had to be selected that represent important attributes of the organisational structure. This was based on the work of Sayer and Walker (1992) who view the effective integration and division of labour as being key to the organisation of industrial production. According to Sayer and Walker (1992) organisational needs occur at different levels within the production system; that is, at the workplace, firm, social relations and spatial structure level. Here emphasis is placed on the intermediate plant level, which combines features of the workplace and firm. Twenty-four binary variables were selected as the basis of the statistical analysis. These refer to firm size and status, production scale and scope, and process continuity and automation.

A Ward cluster analysis based on squared Euclidean distances was the classification method used. At the end of the clustering process, iterative reclassifications were applied according to a hill-climbing algorithm (see Fahrmeir and Hamerle 1984). At first, this method seemed problematic for it normally requires metric data. In the case of binary variables, however, squared Euclidean distances can easily be converted into simple matching coefficients, which are especially designed to measure the similarity of dissimilarity of nominal data. The application of the Ward method was thus justified (Vogel 1975). In the next step, the classification results were compared to those derived using a second approach, which might be better suited to deal with binary data. This was an average linkage procedure based on simple matching coefficients. The results were verified using linear discriminant analysis (Erb 1990). The number of clusters was determined using a heuristic procedure based on the fusion coefficient. In the analysis of the PDPV industry, both classification methods resulted in a 3-cluster solution. Individual firm classifications of both approaches were identical in 74 per cent of all cases. This indicates that the identified organisational clusters represent a relatively stable typology of firms (see Bathelt 1997).

In the final stage of the empirical analysis, the impact of the combined changes in the production environment and the product/process structure on supplier and customer relations was investigated. The questionnaire data, although useful to establish a firm typology, was quite limited in understanding the processes behind changes in production and linkage patterns. It was thus necessary to collect qualitative data to overcome the rigidities of the quantitative analysis. In this stage, a total of six firms (two from each organisational cluster) were chosen as qualitative case studies (for methodological suggestions see Schoenberger 1991, Miles and Huberman 1994). The firms selected had been consistently grouped into the same clusters in the multivariate classification process. The case studies included intensive interviews with executives and managers and analyses of the production technologies and the organisation of labour.[42] The interviews focused on the content, frequency, intensity and spatial aspects of inter-firm communication and adjustment processes.

Organisational clusters in the PDPV industry

In this study, three distinct sets of PDPV firms were identified. The first set consists of 16 semi-flexible, partially integrated firms, the second of 19 flexible, specialised firms and the third of 23 conventional, specialised firms (Table 7.1). These organisational clusters represent a variety of product- and process-related organisational configurations, which have been adapted by PDPV firms to cope with changing technological, economic, institutional and societal settings. It appears that the identified groups represent the most important product and process-related organisational configurations which exist in the German PDPV industry and, as such, a large number of firms within this industry (for a broader discussion, see Bathelt 1997).

Semi-flexible, partially-integrated firms (cluster 1)

This cluster primarily consists of medium sized firms with 100-499 employees, as well as some large firms with 500 or more employees (large producers are not very typical of the PDPV industry) (Table 7.1). Most firms are multi-plant establishments characterised by a network of national and international production facilities. They aim for broad technological or price leadership and produce a wide variety of product groups. In recent years, product changes in this cluster have been rather complex. Differentiation tendencies and horizontal integration in some product groups have been accompanied by standardisation tendencies and vertical disintegration in others. Often, the associated changes have not had a major impact on the overall production programme.

In terms of scale, established processes for large and extremely large batch production are most common in this cluster. Some firms can virtually be classified as mass producers, despite the fact that they employ batch processes. Production technologies used in this group are relatively sophisticated, with a high degree of continuity and automation (for a case study see *Chemie-Produktion* 1997, DAW 1995). Of all groups identified, the firms within this cluster are the most technologically advanced. They apply partially automated, digital process technologies and a wide range of microelectronic devices to support internal planning, training programmes, marketing and distribution activities, material handling procedures and site-specific functional integration. Computerised process technologies are commonly used in conjunction with conventional technologies.

Overall, a partially integrated production structure characterises most firms in this group (i.e. the production of paints and varnishes is usually separated from that of pigments and dyes). The production of high-quality mass products enables these firms to achieve economies of scale. Though many firms have tried to increase product flexibility by producing a larger variety of specialised products, this does not appear to be a central part of their market strategy. Differentiated products are primarily produced to complement and secure markets for standardised products. This is done to strengthen existing links with customers and to block off competitors. The managers interviewed appear to be more concerned about ensuring top quality production than achieving flexibility.

Table 7.1   **Product and process characteristics of PDPV producers by organisational cluster, 1994/95**

Product/process variables	Dominant characteristics		
	Cluster 1 $n = 16$	**Cluster 2** $n = 19$	**Cluster 3** $n = 23$
Firm size	1 medium 2 large	1 small	1 small
Firm status	1 multi-plant firms (headquarters)	1 single-plant firms 2 subsidiaries	1 single-plant firms
Production scope	1 large scope with complex program changes	1 highly specialised scope with horizontal diversif.	1 specialised scope with little changes
Production scale	1 large batches 2 small scale production	1 customised production 2 small batches	1 customised production 2 small/large batch
Degree of process continuity	1 partially- continuous 2 continuous	1 discontinuous 2 partially- continuous	1 discontinuous 2 partially-cont.
Degree of process automation	1 partially-auto	1 partially- auto 2 manual	1 manual
Type of process automation	1 computerised 2 conventional	1 conventional 2 computerised	1 conventional

Note: '1' refers to the dominant, '2' to subordinate characteristics of PDPV producers by cluster and variable.
Source: survey results

Flexible, specialised firms (cluster 2)

The product and process characteristics of cluster 2 firms are quite different from those in cluster 1 (Table 7.1). Almost all producers are small single plant firms or small subsidiaries with less than 100 employees. Highly specialised products dominate their production programme. Standardised products are only of minor importance. Through the production of highly specialised, differentiated products, the firms of this group attempt to become leaders in specialised market segments. Recent production changes have resulted in further horizontal diversification (i.e. the production of a larger variety of products). Due to the specialised product structure, processes for the production of customised products and small batches dominate. Production processes are usually less sophisticated than in cluster 1. They are characterised by conventional technologies with partially automated or manual control. Digital process components are only slowly being introduced.

As a consequence of concentrating on specialised market segments, products are typically designed according to individual customer needs. Since production activities are contract based and extremely small scale, the firms of this group do not need a lot of storage space for their final products. They systematically try to achieve economies of scope, as opposed to scale, based on a flexible, differentiated production programme. Producers of this cluster cope with market segmentation tendencies by opting for distinct flexibility strategies. Despite some obvious parallels, there are clear differences between the PDPV producers of this group and those firms described by Piore and Sabel (1984) in their flexible specialisation scenario. First, the PDPV firms of this organisational cluster are able to achieve a high degree of specialisation and differentiation without further vertical disintegration. Sometimes flexibility even results from integration tendencies. Second, flexible production does not lead to spatial concentration tendencies. Third, flexibility in this cluster is not dependent on the application of advanced process technologies. These producers do not correspond with Conti's (this volume) district firms.

Conventional, specialised firms (cluster 3)

Producers in cluster 3 are predominantly small single plant firms with less than 100 employees (Table 7.1). The product and process characteristics of this group differ substantially from those in the other two clusters. These firms produce standardised as well as specialised products. Their production programme is limited in scope and is aimed towards the achievement of partial market leadership. For this, one of two strategies is employed; that is, either a product-based specialisation strategy directed towards small, well-defined market segments or a spatial specialisation strategy which focuses on near-by customer needs.[43] Even though some firms reported minor differentiation tendencies in their production programmes during the last decade, the production scope of this group has remained largely unchanged.

Production in this cluster is very mixed, including large-scale and small-scale, as well as customised, production. Of the organisational clusters identified, firms from this group employ the least sophisticated process technologies. Conventional production processes with manual control prevail. Digital process controls and other microelectronics applications are rarely used within the group. By using a distinct product based or spatial specialisation strategy, cluster 3 firms try to develop specific problem solving skills, which are not necessarily dependent on the use of advanced process technologies. Attempts to fulfil individual customer needs are only undertaken to obtain larger contracts in the future. Customer-specific designs are not indicative of a systematic flexibility strategy. Customisation activities are carried out to achieve limited economies of scale.

As demonstrated by the identified major product- and process-related organisational structures within the German PDPV industry, a general move towards greater product and process flexibility has clearly not occurred in the last decade.[44] There is also no indication that such a trend can be expected in the near future. Schumann et al (1944) comes to similar conclusions in their study on technological change and labour organisation in the German chemical industry. Only 27 per cent of the plants studied by Schumann et al (1994) employ fully automated digital process technologies. Conventional control engineering still dominates in 48 per cent of the chemical plants. In the remaining 25 per cent, conventional processes are only supplemented by computerised technologies.[45] Since the Schumann et al (1994) study is based only on a survey of some of the most important, large chemical producers in Germany, their results are likely to overestimate the degree of organisational and technological sophistication and flexibility within the industry. Smaller chemical firms are often much slower in adopting flexible machinery and flexible work practices due, in part, to limited finances (see Bathelt 1995b, 1997).

Spatial proximity and the social division of labour in the PDPV industry

Supplier and customer relations in the German PDPV industry have been influenced by changes in the technological, economic, institutional and societal settings and in the product and process structures of the firms. To uncover the dynamics of the buyer-supplier linkages in the industry, the type, frequency and intensity of communication and adjustment processes between producers and their suppliers and customers were analysed. These findings are summarised in the next sections.[46] This includes conclusions regarding the changing nature of the social division of labour and the importance of proximity in industrial relations and practices. The information presented primarily draws upon qualitative data, which was collected during the firm case studies.

Supplier relations

Day-to-day communication and adjustment processes between PDPV producers and their suppliers are fairly standardised. Routine deliveries are usually organised by phone or fax and do not depend on personal interaction or collective learning processes. Nonetheless, PDPV producers are interested in stable, long-term supplier relations. This is because changes in the composition of raw materials and other supplies (such as binding agents, solvents, dyes, pigments, and special additives) can have a significant impact on the characteristics of the final products. Minor alterations in the chemical composition of pigments may, for instance, affect the paints' resistance towards chemicals, heat and humidity or it may change the colour-tone. This would require production alterations to ensure the maintenance of product characteristics and quality. Such adjustments are often associated with significant R&D activities and time consuming testing procedures. Close contacts with suppliers help to avoid problems that may arise when PDPV producers adjust their products to met changes on the supply side. This is especially important for flexible, specialised and conventional, specialised firms since they concentrate their limited R&D capacities towards the fulfilment of customer needs. Their success in certain market segments is based on the ability to formulate specialised solutions in response to non-routine customer specifications. Customisation activities do not allow for internal R&D capacities to be directed to the systematic testing of input materials.[47]

In-depth information exchange between PDPV producers and their suppliers about foreseeable, expected or suggested changes in the composition of raw materials and other supplies takes place in regular meetings. In recent years, meetings have also been an important forum to develop responses to changes in environmental laws and regulations (e.g. the reduction or elimination of heavy metals in pigments and paints). These meetings are often quite institutionalised and result in collaborative development activities and the design of new or varied products.

Relationships of trust are sometimes products of the communication and adjustment processes. They develop over time through face-to-face contacts and experience (similar to the process described by Harrison 1992). Trust between producers and suppliers is important because it helps to undermine the potential threat that plant- and site-specific know-how be transferred to competitors. The danger of such transfers is especially high when suppliers are in contact with a number of different PDPV firms, which compete in the same markets.[48] This does not, however, mean that trust relations will necessarily develop at some point. It is striking that distrust seems to dominate relations between producers and suppliers in some segments of the PDPV industry. In the case of automobiles, producers have caused first tier suppliers (i.e. PDPV firms) to distrust them because of their rigid cost-cutting policies and the creation of fierce price competition in the supplier sector.[49] This, in turn, has stimulated feelings of distrust between first- and second-tier suppliers (i.e. suppliers of PDPV firms). Suppliers of PDPV firms are sometimes not even allowed to enter the production sites for routine deliveries. It is most likely that this has served to restrict opportunities to benefit from learning-by-interacting processes as described by Lundvall (1988) and Gertler (1993).

In the past decade, relations between PDPV producers and their suppliers have expanded and intensified. This is due, in part, to market segmentation tendencies, which have resulted in a diversification of supply needs. In addition, changes in environmental regulations have increased the need for closer producer-supplier collaboration. Overall, this has contributed to a deepening of the social division of labour. Interview results indicate that spatial proximity has been quite advantageous in encouraging producer-supplier co-operation. A number of managers and executives pointed out that proximity promotes trust since private and business contacts overlap. This is supported by the results of the postal survey. Most PDPV firms indicated that they have had long-term, strategically important supplier linkages with nearby suppliers (in a 50 and/or 200 kilometre radius). The resulting linkage pattern has remained fairly stable over the past decade. This does not, however, mean that these same firms acquire most of their supplies within this distance (Bathelt 1997).

Spatial proximity is most important for specialised firms but is also significant for partially integrated producers. One semi-flexible, partially integrated producer has, for instance, attempted to introduce a just-in-time supply system in recent years. The firm offers its suppliers special, long-term contracts to encourage suppliers to move close to the main production site. Thus far, this has had only limited success. The advantages of spatial nearness are especially clear to those PDPV firms who have encountered problems in the past due to a lack of spatial and cultural proximity.[50] In one case, a lack of proximity to a foreign supplier led to a breakdown in communication and resultant production problems. Incomplete information flows about the characteristics of input materials created unwanted changes in the characteristics of the final products. As a consequence, the supplier linkage was terminated. Often, however, firms do not have a choice in ensuring proximity in their supply-side relations since some supplies are only available from a few distant sources.

Overall, the majority of producers primarily contract national suppliers for their inputs (Table 7.2): 10 out of 13 semi-flexible, partially-integrated, 18 out of 19 flexible, specialised and 20 out of 21 conventional, specialised producers acquire their raw materials and other supplies predominantly from German suppliers. Only in the case of semi-flexible, partially integrated firms are linkages with foreign suppliers somewhat stronger (Table 7.2). These firms are commonly integrated into a larger international network of production facilities and have a tradition in dealing with foreign suppliers. The results of the survey indicate that the producers of each organisational cluster have, on average, increased their share of foreign supplies in recent years.

Customer relations

PDPV firms are either user-orientated or trader-orientated in terms of their market relations. Depending on the distribution channels, which they use for their outputs, some of the largest semi-flexible, partially integrated producers to not have direct contact with potential users. Instead, they ship their products to wholesale traders and/or retailers (trader-orientation). In this case, close customer relations are only important to the extent that they help to facilitate long-term production plans and short-term delivery systems. Special communication and adjustment procedures with respect to product quality and characteristics are not required. In other firms, products are designed for specific applications or according to the needs of particular users.

Table 7.2   **Percentage of German and foreign supplies used in the PDPV industry by organisational cluster, 1986/87 and 1994/95**

	Cluster 1		Cluster 2		Cluster 3	
Percentage of supplies	- Number of respondents –					
	1986-87 $n = 13$	1994/5 $n = 13$	1986/87 $n = 19$	1994/95 $n = 19$	1986/87 $n = 21$	1994/95 $n = 21$
	A. German supplies					
Up to 50%	2	3	1	1	1	1
51 to 100%	11	10	18	18	20	20
	B. Foreign supplies					
0%	2	1	6	4	6	5
1 to 20%	3	4	10	11	13	10
21 to 50%	6	6	2	3	1	5
51 to 100%	2	2	1	1	1	1

Source: survey results

These firms produce differentiated paints and varnishes, which are resistant towards abrasion, acids, humidity, etc. (user-orientation). In these markets, a firm's success is dependent on close producer-user collaboration. Flexible, specialised firms produce almost exclusively for individual customer needs.[51] Most batches produced are unique for they represent the output of a special, fine-tuned adjustment process. These firms are dependent on the acquisition of what Cornish (1997) defines as market intelligence. Successful product development is in the case of flexible, specialised firms primarily based on market intelligence acquired through customer contacts in the areas of product testing, sales and services.

Long-term, intensive customer relations develop especially within the user-oriented segment of the PDPV industry because of the need for collaboration. This arises because the adjustment of products according to customer needs is not a routine process. Adjustments can take months or even years and are associated with extensive development and design activities.[52] This requires frequent personal contact and intensive inter-firm communication and creates a basis for collective learning and mutual trust. From a producer perspective, trust is important for it helps to thwart the potential for firm-specific know-how to be transferred to competitors. Sometimes, however, similar to the situation described on the supply side, customers prevent the development of trust. In the production of paints and varnishes for the automobile industry, for instance, a lack of trust has resulted in a highly volatile production environment and has blocked an expansion of the social division of labour.

In recent years, PDPV producers have consciously attempted to intensify and stabilise their relations with customers. Some partially integrated firms have extended their customer/field services to ensure regular contact with important customers. These meetings provide a forum for the resolution of potential problems in the application process and discussion about the production programme. In addition, regular contact with important users helps to identify new customer needs. Another trend in producer-user relations has been an increase in customisation. PDPV producers have used this as a mechanism to stabilise their market position. Furthermore, producers increasingly offer complex products and additional services that cover a range of different, interrelated purposes instead of selling products separately. As part of this, PDPV producers offer problem-solving services, which would have originally fallen within the realm of customer responsibility.

Due to the nature of communication and adjustment processes, spatial proximity can be very advantageous for producer-customer relations. This situation is reflected in the survey results. Most PDPV producers have strategically important customers located within a 50-200 kilometres radius. Conventional, specialised producers have seemingly the strongest local and regional orientation of all firms.

This is also a reflection of the problems that these firms have in approaching customers at greater distances (see Christensen, this volume). Some of the conventional, specialised producers have developed a distinct spatial specialisation strategy and ship the majority of their outputs to customers that are located within a range of 200 kilometres. One of the producers interviewed has an extremely close long-term sales relationship with a customer located only 10 kilometres away. Several times per week, the firm produces high-quality, customised batches to be sent to this customer. Aside from regular producer-user collaboration in the design of new products, both parties meet on a daily basis. The production processes and plans of both firms have been co-ordinated over a long time period to ensure compatibility, an undertaking, which has required significant transaction specific investments.

Table 7.3   **Percentage of German and foreign sales within the PDPV industry by organisational cluster, 1986/87 and 1994/95**

	Cluster 1		Cluster 2		Cluster 3	
Percentage of sales	- Number of respondents -					
	1986-87	1994/5	1986/87	1994/95	1986/87	1994/95
	$n = 14$	$n = 14$	$n = 18$	$n = 19$	$n = 20$	$n = 22$
A. German sales						
Up to 50%	2	3	2	2	-	-
51 to 100%	12	11	16	17	20	22
B. Foreign sales						
0%	4	3	4	3	9	6
1 to 20%	3	4	10	7	8	11
21 to 50%	5	5	3	7	3	5
51 to 100%	2	2	1	2	-	-

Source: survey results

       Flexible, specialised firms have a seemingly weaker local and regional orientation than their conventional, specialised counterparts. Due to the high degree of specialisation, their customer basis is spatially diversified (see also Christensen, this volume). Spatial proximity is thus limited. Often, there are only a few potential customers that are located close-by. In one case, it appeared that a lack of proximity has served to form a barrier for successful product adjustments. In another case, a lack of proximity has made it necessary for a PDPV producer to employ a part-time technician at a customer's site to ensure successful adjustments on a day-to-day basis. According to one of the managers of this producer, this would not have been necessary if both parties had been located close to each other.

As shown in Table 7.3, most firms are strongly oriented towards German customers. This is an indication of the importance of spatial and cultural proximity within the PDPV industry: 11 out of 14 semi-flexible, partially-integrated, 17 out of 19 flexible, specialised and all of the 22 conventional, specialised firms sell more than 50 per cent of their outputs within Germany. Despite the significant role of the national context in terms of unified regulatory, educational and labour relations systems, many PDPV producers have successfully established important relations with foreign customers. Half of the semi-flexible partially integrated and flexible, specialised producers sell at least 20 per cent of their products outside of Germany (Table 7.3). Some of these firms benefit from their international production networks, which allow them to substitute organisational proximity for spatial and cultural proximity. These firms are usually quite experienced in dealing with customers in foreign markets. Overall, a growing number of firms attempt to gain access to or to intensify their activities in foreign markets.[53]

Conclusions: global positioning and regional context

It is premature to conclude that the Fordist crisis has been overcome based on recent economic and societal developments in Germany and other industrialised countries. It is still unclear what changes will lead to a new, consistent mode of development characterised by a new regime of accumulation and mode of regulation. It is not even clear that Fordist practice would first have to be eliminated and replaced before a future mode of development could take hold. This study provides evidence that this, in fact, is not necessary. Despite the limited scope of this study (i.e. a single industry), important directions of contemporary industrial change and tendencies in the technical, social and spatial division of labour can be identified. This is supported by results from the pharmaceutical and basic chemicals industry which have been presented elsewhere (Bathelt 1997,

1999). These results indicate that restructuring activities in large parts of the German chemical industry do not follow a uniform development pattern. Industrial change is clearly more complex than is frequently assumed. Changes in the organisation of production and labour processes are not always related to the Fordist crisis. They also do not inevitably lead to success in resolving the rigidities of the Fordist production structure. It is, therefore, premature to stress notions such as flexibility to the neglect of other dimensions in the characterisation of the newly developing after- or post-Fordist regime of accumulation.

In contrast to what is frequently promulgated, changes made in production programmes and processes are not always meant to follow a single development path towards flexibility. In the PDPV industry, most firms do not fully utilise the potential range of options to increase product and process flexibility. Of all the PDPV producers analysed, only flexible, specialised firms systematically adapt production to meet segmented customer needs to gain economies of scope. Other producers generally do not operate with a distinct flexibility strategy. Instead, most firms try to gain economies of scale through the rationalisation and standardisation of parts of their production programme. New process technologies are often used parallel to conventional technologies; that is, they do not replace them (see also Schumann et al 1994). In the application of new technologies, high quality production is often more important than increased flexibility.

Overall, the product and process structures in the PDPV industry and other chemical branches indicate a high degree of persistency (Bathelt 1997, 1999). Iterative refinements in production programmes and processes are more dominant than radical shifts in response to new technological, economic, societal and institutionally settings. The results of this survey do not support Freeman's (1990) hypothesis that a general shift in the techno-economic paradigm from energy- and material-intensive, standardised mass production towards information-intensive, flexible product and process configurations is taking place in the chemical industry. Furthermore, restructuring activities from the production of mass products towards speciality chemicals have a long tradition in the chemical industry and are, thus, not necessarily related to the Fordist crisis (Schall 1959).

The social division of labour also appears to remain relatively fixed. Survey results do not provide evidence that a drastic shift in buyer-supplier relations is taking place in the PDPV industry. Most firms try to establish stable, long-term supplier and customer linkages. Flexible, specialised and conventional, specialised firms rely particularly on close supplier contacts since they direct their R&D activities primarily towards fulfilling customer needs. Frequent information exchange with suppliers helps to avoid problems in the adjustment of products and processes to account for changes in the composition of raw materials and other supplies. Over time, such contacts help to stimulate trust. Stability and continuity in supplier linkages increase the efficiency of communication and adjustment processes. Similar relations also exist on the customer side. User-oriented PDPV firms try to develop close long-term customer linkages. This is important in the adjustment of products to meet particular quality standards, to identify changing customer needs and enter new market segments. Producer-user relationships are generally stable because a change of suppliers would be quite costly from the view of the customer (see Christensen, this volume). Flexible, specialised PDPV firms are most strongly oriented to having closer customer relations since they focus on customised products and individual product designs.

It is also not yet clear which spatial patterns will dominate the organisation of industrial production in an after- or post-Fordist mode of development. The question is still open as to whether industrial agglomerations will retain their ability to advance economic growth at a local and regional level or whether globalisation tendencies in production will reinforce decentralisation and deconcentration processes. As suggested by the survey results presented above, which reflect Storper's (1997) line of argument, spatial proximity can be beneficial in different ways in the communication and adjustment processes between chemical producers and their suppliers and customers. In opposition to Krugman (1991), the analysis presented here reveals that technological spillover effects are clearly important in the PDPV industry. Therefore, it is not surprising that most firms in this and other chemical branches have strategically important suppliers and customers located within a 50-200 kilometre radius (Bathelt 1997). This should not be interpreted to mean that these firms acquire most of their supplies and sell most of their final products within this zone.

In response to the changing economic, technological, institutional and societal settings of the Fordist crisis, PDPV producers and other chemical firms employ more complex programme and process configurations. As Conti (this volume) points out, the contemporary economy is based on a variety of structures and the exploration of different development paths. According to the results of this study, Fordist organisational structures are still in place. Their disappearance has thus been largely over-emphasised in the literature. In fact, Fordist practices in some markets are seemingly more successful than ever. At the same time, modern process technologies and organisational principles increasingly supplement Fordist structures and allow for a limited increase in product and/or process flexibility. Overall, a hybrid production structure seems to be developing in the PDPV industry and other chemical branches; one which contains both elements of rigidity and flexibility. Semi-flexible production structures are concentrated in volatile markets, while experience-based production and traditional technologies are still prevalent in stable markets. Large integrated firms and small and medium sized specialised firms are found in both market types but often concentrate their activities in different product segments. The survey results do not support the hypothesis that large firms generally attempt to push small and medium sized firms out of the market or that, alternatively, the latter group would gain in importance as Fordist rigidities are overcome. In the PDPV industry, both groups seem to coexist and tolerate one another, albeit without intensive co-operation.

A number of firms indicated one trend, which could become a potential problem in the future. Large integrated suppliers and customers increasingly use their organisational capacities to adjust to changes in global markets and competition. By establishing international production and distribution networks, a stepwise reduction of activities at their former production sites results. This, in turn, negatively impacts small- and medium sized, specialised PDPV producers because it limits their ability to establish relations with near-by suppliers and customers.[54] Overall, the risks associated with the social division of labour might increase over time and PDPV firms may become more vulnerable to further changes in their economic, technological, societal and institutional settings. At this point, however, Germany is not very likely to lose its status as a geographical centre of PDPV and other chemical production. This is because the spatial development of these industries is subject to pre-existing industrial agglomerations and the localised capabilities, which have evolved therein.

PART III

LOCALISED RESOURCES AND LOCALISED LEARNING: A COMPETITIVE ADVANTAGE?

8 Localised Knowledge, Interactive Learning and Innovation: Between Regional Networks and Global Corporations[55]

Bjørn T. Asheim and Arne Isaksen

Introduction

This chapter examines how firms in three regional clusters in Norway – the shipbuilding industry at Sunnmøre, the mechanical engineering industry at Jæren and the electronics industry at Horten – exploit *both* place-specific local resources and institutions as well as external, world-class knowledge, respectively to strengthen their competitiveness. From these case studies, we make five points:

- ideal-typical regional innovation systems, i.e. regional clusters 'surrounded' by supporting local organisations, are rather uncommon, at least in a country like Norway;
- external contacts, outside the local industrial milieu, are crucial to innovation processes in many SMEs;
- innovation processes may, nevertheless, be regarded as regional phenomena in regional clusters, as regional resources and collaborative networks often have decisive significance for firms' innovation activity;
- regional resources include both tacit and codified knowledge that, combined, are place-specific, contextual and geographically immobile;

- innovation activity in the three regional clusters can be conceptualised as 'regional innovation networks'.

Regional clusters between globalisation and regionalisation

The globalising world economy is characterised by two partly contradictory tendencies. On the one hand, we can identify the *neo-Fordist* development path of world-wide sourcing based on the principle of *comparative advantage* and the lowest possible input costs (i.e. the best relative access to, and most efficient use of, 'natural' production factors). This development has been facilitated by developments in transportation and communication technologies, and by the liberalisation and de-regulation of international trade and financial markets. On the other hand, we have the *post-Fordist* development path of the learning economy, in which global competition is based on the more dynamic principle of *competitive advantage*, resting on "making more productive use of inputs, which requires continual innovation" (Porter 1998, p.78). Continuous innovation in a learning economy is conceptualised as a localised interactive learning process, which is promoted by clustering, networking and inter-firm co-operation, and organised as outsourcing and vertical disintegration so that companies can cut costs and increase flexibility in unstable markets. Companies have trimmed back their in-house operations to core competencies and out-sourced other activities to formally independent suppliers (Harrison 1994). Even if both tendencies are constituted by, and constitute, the globalisation process, the neo-Fordist development trajectory is often referred to as *globalisation*, while the post-Fordist development path is described as regionalisation.

As discussed in Chapters 4 and 5, economic globalisation means increased interdependence between firms and production systems in different nations and regions. Regional and national supplier and sub-contracting systems are increasingly integral to global production networks and systems, and are thus exposed to international economic conditions (Andersen and Christensen 1998). The debate on globalisation suggests that national and regional authorities may lose their influence over industrial development in their territories as firms are incorporated in global commodity chains run by TNCs. Decisions on downsizing, closure and relocation are taken directly (for plants owned by TNCs) and indirectly (for suppliers) by decision makers in remote corporate headquarters and not by local entrepreneurs. As firms are tied into evolving international organisational structures, the continuous reorganisation of 'global firms' has the capacity to reshape dramatically the fortunes of regional economies.

Besides, economic globalisation is seen as making formerly locally embedded production factors globally available, eroding previously advantageous local factors of production, and exposing firms to international (price) competition (Maskell et al 1998).

Regionalisation can also in a direct way form part of the globalisation process, as some innovative, regional clusters are seen to play an important role in global production networks. As it was put by Amin and Thrift (1992), "they represent neo-Marshallian nodes in global networks. These nodes act as 'centres of excellence' in a given industry, offering for collective consumption local contact networks, knowledge structures and a plethora of institutions underwriting individual entrepreneurship" (p. 577). Thus, "corporate globalisation strategies are meaningful only if local, national and regional differences exist and can be harnessed on a global scale" (Braczyk and Heidenreich 1998, p. 414). In some localities, global firms may find access to specialised, experience-based and place specific knowledge, a flexible regional production system, and a soft infrastructure of enterprise support provided by innovative regional governance institutions striving to create competitive local environments. Then, "the creation of regional clusters and the globalisation of production go hand in hand, as firms reinforce the dynamism of their own localities by linking them to similar regional clusters elsewhere" (Saxenian 1994, p. 5). Thus, foreign direct investment is often attracted to specific regional locations. In some cases, this may result in 'embedded' branch plants - plants with some decision-making autonomy that base some activities on place-specific knowledge and other assets to help them in their battles for intra-corporate contracts and investments. The regionalisation process points to the fact that innovative regional clusters may provide the best basis for achieving competitiveness in the global economy, a position often neglected in the globalisation debate, focused on the homogenisation of factors of production and global sourcing strategies.

While globalisation may be seen as 'the end of geography', as global economic integration removes geographical boundaries, the parallel process of regionalisation points to an increased importance of place-specific and often non-economic factors in creating competitive advantage and differences in regional economic growth rate. Thus, Porter argues that "the enduring competitive advantages in a global economy lie increasingly in local things – knowledge, relationships, motivations – that distant rivals cannot match" (Porter 1998, p.78). Regionalisation refers to economic activity dependent on resources which are specific to individual places, and which cannot easily or rapidly be created or imitated in places that lack them (Storper 1997). Regionalisation emphasises the way systems of firms are embedded in local economic, social and cultural structures that are important for firms' competitiveness

The subject explored in this chapter is how firms in regional clusters[56] may be influenced by, and themselves influence, globalisation and regionalisation processes. The aim is to investigate the usefulness of a multilevel approach to innovation systems for promoting regional clusters, as firms' innovation activities come to rely both on place-specific and experience-based resources and competencies, as well as formally codified knowledge available at the national and global level. The chapter is divided into four parts. In the first part, the key argument of the regionalisation thesis is summarised as the basis for promoting competitiveness. In the second part regional innovation systems are defined and some of the limitations of the regionalisation thesis for analysing regional clusters SMEs are highlighted. The third part illustrates the theoretical arguments through analyses of how firms in three regional clusters in Norway exploit the geographical scale and scope of knowledge infrastructures and innovation systems to strengthen their competitive advantage. Finally, in the fourth part, some theoretical and empirical conclusions are drawn from the three case studies, pointing to the need to reconsider some elements of the regionalisation thesis.

The regionalisation thesis: the creation of competitiveness in regional clusters

The crux of the regionalisation argument is that the regional level, often meaning local labour market areas and specific local and regional resources, may still be important as firms strain to gain global competitiveness, despite economic globalisation. A wide range of literature has emphasised regionalisation as at least a partial solution to understanding dynamic industrial development in some places as well as solving regional economic development dilemmas stemming from the new competition in the globalised economy (Pike and Tomaney 1999). Two empirical observations in particular point to a link between, on the one hand, geographical location and place-specific resources, and, on the other hand, industrial performance (Malmberg et al 1996). The first is that entrepreneurial and innovative activities in some industries tend to agglomerate at certain places. Although based on anecdotal evidence, some places are especially dynamic and innovative, and often within a particular industrial sector or production system. These places have 'something' which is special for creating dynamism not found in other places. Often omitted in the literature on 'regional innovation' is the continuing capacity of a regional economy to create new firms to secure its long-term vitality. New firms, and especially entrepreneurial firms, are partly the results of innovative activity in existing firms. However, the regional ability to form new firms may also rest on locally specific norms and cultural patterns that encourage or impede people to create new firms. It may also rest on an institutional environment that reduces the difficulties of launching new firms and stimulates early growth. That environment might include venture capital funds and local expertise on legal and financial matters (Kenney and von Burg 1999). Thus, "today's economic map of the world is dominated by ... clusters: critical masses – in one place – of unusual competitive success in a particular field" (Porter 1998, p.78).

The second observation is that the factors creating dynamism and competitiveness in successful firms seem to be determined by conditions in the firms' environment, where environment should be understood both in a territorial and in a functional sense. In a territorial sense, conditions in the local industrial milieu, such as formal and informal knowledge, other specialised firms, informal institutions that co-ordinate information exchange and collaboration, entrepreneurial attitudes and networking organisations, seem to be of great importance for firms' competitiveness. In a functional sense, firms' contact and co-operation with other firms and organisations, independent of geographical location, are of great importance. Firms may, however, exploit different types of resources in their territorial as opposed to their functional environment.

What are the mechanism behind the dynamism and competitiveness found in some regional clusters? The regionalisation thesis has four main building blocks:

- *innovation* is increasingly seen as a way for firms, regions and nations to gain competitiveness in the face of globalisation because it enhances the learning abilities of firms and workers (Lundvall and Borrás 1997);
- innovation is conceptualised as an *interactive learning* process, emphasising the importance of co-operation and mutual trust in promoting competitiveness, further promoted by proximity;
- learning is seen as mainly a *localised* process, pointing to the importance of historical trajectories and 'disembodied knowledge', which are highly immobile in geographical terms; and
- as a consequence, agglomeration and *regional clusters* are looked upon as an efficient basis for interactive learning, arguing for the importance of bottom-up, interactive regional innovation systems and networks specifically, as well as untraded interdependencies in general, as a stimulus for innovation and learning.

Innovation

The *first* building block in the regionalisation thesis is the by now familiar conceptualisation of the contemporary post-Fordist economy by Lundvall as a globalising learning economy. "Globalisation has not only increased market competition, but also transformed it into market competition based increasingly on knowledge and learning" (Lundvall and Borrás 1997, p.28). However, the concept of the learning economy can be used in two interconnected ways; as a theoretical perspective on the economy, and as a reference to the current, historically specific period in which knowledge, innovation and learning have attained greater importance in the economy. It therefore requires a new theoretical framework; other than standard economic theory, for it to be analysed (Lundvall 1996). 'Strong' competitive strategies and innovativeness are seen to be inextricably linked. While capitalism has always rested on its capacity to create new products and new ways of producing them (Hudson 1999), a common assumption is that the contemporary economy is less standardised and predictable than it was in the heyday of Fordism, requiring innovation and adaptation to be competitive. Thus, it is the capacity to learn and to innovate that increasingly is seen to determine the relative successfulness of firms, regions and countries. The increasing importance attached to knowledge assets and learning abilities for building the competitiveness of local, regional or national industrial milieus adds these as new important production and location factors influencing the geographical pattern of industry (Malmberg 1997). They are, therefore, new and important issues that need to be understood in analyses of regional uneven development.

Interactive learning

The *second* building block in the regionalisation thesis is the view of innovation as a complex, interactive, non-linear learning process, where learning includes the building of new competencies and establishing new skills among workers and firms, and not only access to new information. This complements the first building block in which innovation involved the introduction of new things and new ways of working by putting new knowledge into use or by using existing knowledge in new ways. This second view of the innovation process is based on a broad definition of innovation to include both improvements in technology and better methods or ways of doing things. It can emerge as new or changed products, services and production methods, new approaches to marketing, new forms of distribution and changes in management, work organisation and skills of the workforce (EC 1995). This view involves a critique of the linear, sequential model of innovation, which focuses on more radical technological innovations. However, "much innovation, in practice, is rather mundane and incremental rather than radical. It results from organisational learning as much as from formal R&D" (Porter 1990, p. 45). These arguments, together with the broad understanding of innovation, mean that the range of innovative industries must be extended beyond the typical centrally-located high-tech industries to include traditional, non-R&D-intensive industries often located in peripheral regions.

The conceptualisation of innovation as interactive learning emphasises the importance of co-operation in innovation processes as well as a systemic view of innovation. The concept of the innovation *system* is thus based on the idea that the overall innovative performance of an economy depends to a large extent on how firms use the experience and knowledge in other firms, research institutions, on government sectors, and mix this with their internal innovative capabilities (Gregersen and Johnson 1997). Thus, the build-up of different local organisations to create 'institutional thickness' (Amin and Thrift 1994) is emphasised as important in stimulating co-operation, learning and innovative activity. Firms combine resources and knowledge from many actors and build unique, firm-specific competencies that cannot rapidly be imitated by competitors (Maskell et al 1998). Entrepreneurs and firms that fail to learn and change their products and ways of doing things will sooner or later exit from the market.

With this perspective on innovation networking, interactive learning and co-operation are considered to be of strategic importance in promoting competitiveness in the post-Fordist learning economy. These are completely different forms of economic co-ordination from those prevailing under Fordism, which was based more on hierarchical control than on trustful co-operation (Cooke 1998). Customer firms have lost some elements of their hierarchical power by relying on their suppliers' creativity as a foundation for their own improvements, although the networks involves still have important power asymmetries. A tendency that has been observed is for firms to develop close, long-term co-operation with suppliers - relationships that may also extend to innovation (Isaksen et al 1999). Quality, reliable delivery, rapid response to complaints, personal relationships and the possibility of co-operation on innovation projects using suppliers' expertise are all important factors behind closer relationships. Two important aspects of the network paradigm seem to be of importance for knowledge accumulation and diffusion: 'economic gains' from networking, and a social aspects of the workings of networks, respectively. The first aspect points to the fact that the external division of labour within networks of firms may lead to the accumulation and creation of new knowledge through *extended specialisation* at the firm level. Firm-level specialisation in networks allows firms to concentrate on core activities while letting specialised sub-contractors carry out complementary activities. This kind of specialisation can lead to high levels of competence within groups of firms in relatively narrow fields. Network increases the chances of new product solutions, and new cost-efficient production methods being identified increasing aggregate competitiveness.

The second aspect relates to the importance of informal institutions in shaping learning processes and entrepreneurial activities. Innovation involves communication, and the flows of information are almost always interpersonal, human linkages. These linkages are quite different from arms-length, anonymous market transactions, and the existence of informal institutions facilitates collaboration and the exchange of qualitative information between actors. Thus, "in networks and other kinds of "organised" market relations, people develop codes of communication, styles of behaviour, trust, methods of co-operation etc. to facilitate and support interactive learning" (Gregersen and Johnson 1997, p. 482). These kinds of social institutions mean that firms may co-operate without written contracts, as people know and follow the same established practices, routines and unwritten rules of business behaviour and rely upon trust in relationships. For Silicon Valley, Saxenian (1994) describes these characteristics as "community ..., with a shared language and shared meanings ... distinguished by the speed with which technical skill and know-how diffused within a localised industrial community" (Saxenian 1994, p. 37). All this points to the importance of the socio-cultural milieu within which networks of firms are embedded, where innovations are the result of social interaction between economic actors.

Localised learning

The *third* building block of the regionalisation thesis begins with innovation as a social process influenced by its cultural and institutional contexts. This also implies that learning is to a large extent *localised* and not placeless. This view gives "particular emphasis to history, routines, influences of environment and institutions" (Cooke 1998, p. 7). The competitiveness of firms is partly seen to depend upon the stock of knowledge and the learning ability in the regional milieu.

To explain why learning is often a localised process, it is useful to make the distinction between two main types of knowledge; formal, codified (scientific or engineering) knowledge and informal, tacit knowledge, which is constituted by skilled personal routines, technical practises and co-operative relations. Codified knowledge corresponds mainly with Lundvall and Johnson's (1994) concepts of 'know-what' ('facts') and 'know-why' (scientific principles). Tacit knowledge corresponds with 'know-how' (personal skills) and 'know-who' (information about who has specialised knowledge). As a rule, codified knowledge is created through systematic research activities in R&D institutions, universities etc. This knowledge is, in principle, universally available. However, to acquire the knowledge requires contact with knowledge institutions as well as an existing stock of codified knowledge. Firms obtain 'know-why' knowledge by recruiting educated workers and through direct contact and co-operation with knowledge institutions. Additionally, much knowledge flows are embodied in physical capital like materials, machinery, components and products. This kind of knowledge is often 'invisible' to its users, but is nevertheless employed indirectly (Nås 1998).

Tacit knowledge is part of human and social capital. This type of knowledge consists of individual skills of a more or less intuitive kind (know-how) that workers must learn in their 'daily' work (i.e. learning by doing and using), and often in interaction with more experienced colleagues, in apprenticeship-like relationships. Tacit knowledge is also anchored in specific routines, norms of behaviour, implicit and shared beliefs and modes of interpretation stimulating communication in organisations, firm networks and local communities (Lundvall and Borrás 1997). Know-how cannot easily be isolated from its individual, social and territorial contexts. The 'know-who' form of tacit knowledge is also seen as socially embedded knowledge that is difficult to codify and transfer through formal channels of information. Thus, tacit knowledge is difficult to transfer because it is not an explicit form of information, and then it can not be traded as a commodity. The labour market, however, is partly a market for know-how, where firms compete to recruit people with particular informal (and formal) skills. Still, tacit knowledge is often firm and industry specific, as well as "embedded in a multitude of inter-firm relationships and therefore cannot be taken out of context without losing much of its value" (Malmberg et al 1996, p. 92). Some kinds of networking emerge precisely because firms need to gain access to the tacit knowledge of other firms in order to achieve synergy with their own knowledge and skill.

Maskell et al (1998) claim that knowledge is increasingly being codified. The process of codification makes tacit and locally embedded knowledge globally available or turned into a 'ubiquity' (i.e. knowledge is transferred from Box III to Box II in Figure 8.1). When this happens a local comparative advantage disappears. It is argued that codification may undermine the competitiveness of firms in the high-cost areas of the world.

Firms in low-cost regions become becoming increasingly competitive when, in addition to the advantage of low costs, they are able to utilise the most efficient production technologies, the latest organisational designs and so on.

	Local 'sticky' knowledge	Global 'ubiquitous' knowledge
Formal, codified knowledge ('Know-what' and 'know-why')	I. Scientific knowledge which is locally disembodied as it is produced in co-operation between local firms and R&D-institutions, and since some tacit knowledge is required in order to handle and use the codified knowledge	II. Scientific knowledge and information produced in R&D institutions and universities, that may be transferred through formal learning, recruiting, textbooks, manuals, and via the purchase of machinery, component etc.
Informal, experience based, tacit knowledge ('know-how' and 'know-who')	III. Firm-specific knowledge and knowledge produced in networks of (often co-located) firms, produced and transferred through learning by doing, by using and by watching	IV. Knowledge which can be transferred through recruiting workers with specific experiences, but the knowledge may be of less value and difficult to use outside its local context or outside specific firms and industries

Figure 8.1   Classification of different kinds of knowledge

Nevertheless, high cost areas often maintain competitive advantage over long periods. This persistence can be explained by the localised and 'sticky' nature of some forms of knowledge and learning processes. Thus, the formation of and access to tacit knowledge and knowledge shared by people in organisations and in inter-firm networks and co-operative arrangements is now seen as the key to economic success. "[M]ost economically useful kinds of knowledge have a tacit dimension and ... such knowledge can only be obtained in a social process of interaction" (Lundvall and Barrás 1997, p. 50). Some places are favourable for the development and diffusion of this kind of tacit knowledge, in particular industrial milieus with a long tradition in some specific industries and technologies and characterised by close co-operation and mutual trust. Even if tacit knowledge is being codified, firms may still profit by being located in places where tacit knowledge is constantly created and maintained. Then it is vital to view innovation also as a 'bottom-up' process, and to capture the knowledge and creativity of all (core) workers to be engaged in step-by-step improvements of products (or services) and productivity.

However, it is not possible to codify, and make ubiquitous, all kinds of tacit knowledge (i.e. not all tacit knowledge may be transferred from Box III to Box II in Figure 8.1). Some knowledge is internal to people and organisations, and exists as routines, habits and informal procedures in firms. It is shared only among people in local communities. Normally this knowledge has to be acquired by on-the-job training, learning by doing and by using and by entering into processes of interactive learning. If it is the case that codified knowledge becomes ubiquitous and is put into use by firms in many places at an increasing speed, the importance of tacit and non-codifiable knowledge increases. This is knowledge that cannot be copied quickly and easily by competitors, but the knowledge may be important for the innovative ability in firms, and thus constitutes a local competitive advantage. Thus, "in a knowledge-based economy, agglomerative benefits spring from the need to access knowledge which cannot easily be acquired on the market" (Maskell et al 1998, p.62). The quality of knowledge has become one of the few production factors to vary spatially.

We argue here, however, that an important part of codified knowledge is, like tacit knowledge, the result of localised learning and is geographical immobile. Thus, codified *and* contextual disembodied knowledge, which is *not* embodied in machinery, and which can be both tacit and codified (Box I, Figure 8.1), may be an important resource for firms to generate radical innovations. Firms then build their localised learning on a strategic use of codified, R&D-based knowledge, in addition to tacit knowledge. This knowledge is 'sticky' since it may be developed through collaboration between, for example, local R&D institutions, technology centres and firms. The knowledge is partly embedded in local patterns of interaction and in 'know-who'. The local area contains people with first-hand operational experience of that knowledge. The adaptability of this localised form of codified knowledge is dependent upon and limited by, contextual, tacit knowledge (Asheim and Cooke 1998). Generally speaking, the relationship between tacit and codified knowledge is seen to be symbiotic, as tacit knowledge may be necessary to handle and use new, codified knowledge (Lundvall 1996).

The strict dichotomy normally applied between codified and tacit knowledge can, thus, be quite misleading both from a theoretical and policy point of view especially if localised learning is primarily said to be based on tacit and practical knowledge. A claim for the superiority of tacit knowledge on such grounds holds the danger of fetishising its potential to make regional clusters globally competitive. It ignores the problems firms in some clusters face owing to their lack of strategic, goal-orientated actions and strategies, which, necessarily, have to be supported by codified, R&D-based knowledge (Amin and Cohendet 1997). Institutionalisation of tacit knowledge in the routines and operating procedures of organisations may lead to incremental innovation along dominant technological and regional trajectories, maintaining competitiveness in periods of relative stability. While in danger of being reductionist, a technology's trajectory and potential may be seen as vital for understanding the fates of regional industries based on a dominant technology (Kenney and von Burg 1999, p.68). Thus, a supplementary explanation behind the different development of Silicon Valley and Boston Route 128 could be the path-dependent technological trajectories of their leading industries. Saxenian (1994), in her already classic book on the two regions, focuses more on the different cultures and forms of organisation of the two industrial systems: Silicon Valley becoming the centre of the semiconductor industry, and Route 128 the centre of minicomputer production. While semiconductors are a fundamental input to all electronics products, the minicomputer is more limited and has less market potential.

The down-side of cumulative learning and path-dependency is institutional, social and cultural 'lock-in', where a local production system, for example, upgrades knowledge and products in a branch of industry that is becoming technologically anachronistic (Cooke 1998). Thus, regional clusters that relay mainly on tacit knowledge may experience difficulties in periods of instability, when a rapid response through radical innovation may be required. More generally, it is doubtful whether incremental innovation, based on tacit knowledge, will be sufficient to secure the necessary long-term competitiveness of regional clusters in an increasingly globalised world.

When local disembodied knowledge, both tacit and codified, are significant in advancing innovative capacities and the competitiveness of firms – and are of increasing significance as knowledge is codified and made globally available – firms may profit by having access to this kind of geographical immobile knowledge. The knowledge assets and learning capacities of particular local, regional or national milieus are an important competitive factor in the current economy. Thus, "one of the few remaining genuinely localised phenomena in this increasingly 'slippery' global space economy is precisely the 'stickiness' of some form of knowledge and learning processes" (Malmberg 1997, p.574). The best way for firms to acquire this 'sticky' knowledge is to be located in areas (often, dynamic regional clusters) where learning processes develop new and economically useful knowledge. By having a plant, R&D department, supplier or collaborator in such a place, the knowledge base of such a region can be tapped. This occurs "when the content of knowledge is changing rapidly ... and only those who take part in its creation can get access to it" (Lundvall and Borrás 1997, p.34). Regional collective learning occurs in some particular places involving "the creation and further development of a base of common or shared knowledge among individuals making up a productive system" (Keeble and Wilkinson 1999, p.296). A base of shared knowledge and reciprocity between actors in a regional cluster facilitates the combination of diverse knowledge to generate new knowledge.

Cluster

The *fourth* building block in the regionalisation thesis concerns agglomerations, and in particular regional clusters, as places where close inter-firm communication, socio-cultural structures and institutional environment may stimulate socially and territorially embedded collective learning and continuous innovation. The crux of the argument is that "the proximity between different actors makes it possible for them to create, acquire, accumulate and utilise knowledge a little faster than their cost-wise more favourable located competitors" (Maskell et al 1998, p.59). Thus, in essence "the theoretical case for territorially based systems of innovation builds on interactive learning" (Gregersen and Johnson 1997, p.482).

In the case of regional clusters, localisation economies refer to the co-presence of many firms in the same or adjacent industrial sector, giving rise to economic benefits accruing from proximity. Firms may achieve cost reductions from having access to pools of common factors of production in the region and opportunities to collaborate in bulk purchasing, that are "a variety of external economies associated with location" (Harrison 1992, p. 472). Lower unit costs of production are achieved because firms build up and have good access to *common* input factors such as skilled labour, specialised service firms, various types of technical infrastructure, and other localised externalities. Firms may join forces, and, for example, co-operate with regional and municipal governments, to overcome common bottlenecks in infrastructure, labour supply and so on.

The Marshallian understanding of agglomeration economies depicts the more qualitative and social aspects associated with a territorial concentration of industrial production. It points to the fact that the competitiveness of regional clusters lies in their superior ability to enhance information sharing, collective learning, flexible adjustment, and the chance to pick up ideas from the early adopters of new technologies, for example. This opportunity is subtle and is, to a large extent, social rather than purely economic (Malmberg et al 1996). Explanations of cluster dynamics have increasingly turned away from economic reasons, such as localisation economies. Instead, they have turned to social-cultural reasons, "such as intense levels of inter-firm collaborations; a strong sense of common industrial purpose; social consensus; extensive institutional support for local business; and structures encouraging innovation, skill formation, and the circulation of ideas" (Amin and Thrift 1994, p.12). These are untraded interdependencies, i.e. "a structured set of technological externalities which can be a collective asset of groups of firms/industries within countries/regions and which represent country- or region-specific 'context conditions' of fundamental importance to the innovative process" (Dosi 1988, p.226). Trust as a fundamental aspect of social capital in interactive learning may be viewed as a localised, non-tradable business input, because trust cannot be bought, and even if it could be, it would be of little value. Thus, a regional capability, rooted in particular patterns of inter-firm networking and inter-personal connections cannot be transferred to other places. "It can only be built up over time" (Lawson and Lorenz 1999, p.10).

The Marshallian view of agglomeration economies predates the idea of 'embeddedness' (Granovetter 1985) as a key analytical concept in understanding the working of some regional clusters (i.e. industrial districts). It is embeddedness in broader socio-cultural factors that is the material basis for Marshall's view of agglomeration economies. Marshall emphasised in particular 'the mutual knowledge and trust' that reduces transaction costs in local production systems; the 'industrial atmosphere' that facilitates the generation of skills and qualifications required by local industry; and the combined effects of these aspects in promoting (incremental) innovations and innovation diffusion among SMEs in industrial districts (Asheim 1992).

Analytically, the specific *local* conditions in regional clusters underlying dynamism and innovative ability may be specified as traditional localisation economies, territorially embedded Marshallian agglomeration economies, untraded interdependencies and disembodied knowledge. The combination of these factors could constitute the material basis for a new form of socially created competitive advantage for regions in the globalised economy.

Regional innovation systems – some limitations of the regionalisation thesis

The new understanding of innovation activity and regional clusters has focused attention on the term 'regional innovation system'. Regions are seen as important bases of economic co-ordination at the meso-level. "The region is increasingly the level at which innovation is produced through regional networks of innovators, local clusters and the cross-fertilising effects of research institutions" (Lundvall and Borrás 1997, p.39). Thus, a strong case is made today for regional clusters and innovation systems being of growing in importance as a mode of economic co-ordination in the post-Fordist learning economy. Close co-operation with suppliers, sub-contractors, customers and support institutions in the region will enhance the process of interactive learning and create an innovative milieu favourable to innovation and constant improvement. This argument corresponds with Porter's (1998) emphasising of clusters as critical for companies' 'home base' activities that create and renew their products, processes and services – and where the innovation potential of clusters are the critical resource.

The 'regional innovation systems' (RIS) is a theoretical construct to analyse and come to grips with the working of regional clusters. It highlights actual development tendencies in the building of networked innovation architectures in regions. It is also a policy tool for the creation a supportive system of innovation at a regional scale (Cooke 1998).

To delimit RIS, it is first useful precisely to define what is meant by 'regional clusters'. Here we follow Rosenfeld (1997) who defines a regional cluster as a geographically bounded concentration of *interdependent* businesses. Rosenfeld stresses that clusters should have active channels of business transactions, dialogue and communication. "Without active channels even a critical mass of related firms is not a local production or social system and therefore does not operate as a cluster" (Rosenfeld 1997, p.10). This definition uncovers two main criteria for delimiting regional clusters.

First, regional clusters are limited geographical areas with a relatively large number of firms and employees in one or a few related industries. They are *specialised*. Second, although firms in regional clusters may co-operate with firms, R&D institutes and so on in many places, the firms are part of *local networks*, which are often production systems. However, the firms may be interlinked in other ways, for example by the use of a common knowledge base, the same raw materials, or by basing their interactions on social values and collective visions that foster trust and reciprocity. The geographical extent of a cluster is a function of 'place'. They may embrace local labour markets or travel-to-work areas, for example.

Both Cooke (1998) and Porter (1998) employ a wider definition of regional clusters, including different kinds of formal institutions as supporting their innovative capacities. In Porters words, "clusters are geographic concentrations of interconnected companies and *institutions* in a particular field" (p.78, italic added). Here we favour a more geographically restricted definition of interconnected *firms* to identify clusters, and use the concept of the *regional innovation system* to denote regional clusters surrounded by 'supporting' institutions. Basically, regional innovation system comprises two main types of interacting actors (Asheim and Isaksen 1997). The first type is the firms in the region that make up a local or regional production system. The second type comprise the local institutional infrastructure that holds important competencies which support regional innovation: i.e. research and higher education institutes, technology transfer agencies, vocational training organisations, business associations, finance institutions. Regional clusters are mainly *spontaneous* phenomena; geographic concentrations of firms often developed through local spin-offs and entrepreneurial activity. "Once a cluster begins to form, a self-reinforcing cycle promotes its growth" (Porter 1998, p.84). RIS often have a more *planned* and systemic character. Thus, development from a cluster to an innovation system requires a strengthening of the local institutional infrastructure that supports innovation co-operation. The objective is often to increase innovation activity and capability in a region by the use of policy tools.

Brusco's (1990) division of Italian industrial districts[57] into Mark I and Mark II types illustrates the analytical potential of distinguishing between regional clusters, networks and RIS. Mark I districts have few or no local, public policy tools to stimulate innovation activity among local firms. The local system of firms is nevertheless innovative, principally through incremental product and process innovation. Innovative activity builds upon the experience-based competence of entrepreneurs and skilled workers and close local co-operation in an atmosphere of mutual trust and understanding.

During the 1980s, an important task was to develop some districts into Mark II districts to increase their innovative capacities beyond incremental innovation. This was necessary to counteract competition from low-cost countries and to introduce more advanced technology requiring more radical innovation. Up-grading districts to Mark II status occurred through, in particular, the establishment of stronger, local, institutional infrastructure, providing real services with specialised competencies (market development, technology, strategy etc.) adapted to the dominant sector in the district. These service centres supply their local systems of firms with professional competencies that small firms seldom acquire themselves, but which are often necessary to accomplish radical change. Generally, contemporary analyses of industrial districts emphasise the collectivist and institutional basis of successful co-ordination (Keeble and

Wilkinson 1999). This development of industrial districts from Mark I to Mark II status, or from informal to formal networks, is an important stimulus for policy-makers, suggesting a way of increasing innovativeness in regional clusters through the use of dedicated policy instruments.

What this extension of the definition of the concept of cluster also indicates is a deepening and widening of the degree and form of co-operation taking place within clusters. The original and simplest form of co-operation within a cluster can often be described as *territorial* integrated input-output (value chain) relations. These could be supported by informal, social networking, as is the case with Marshallian agglomeration economies, but they could also take the form of arms-length market transactions between a capacity sub-contractor and a client firm. The next step of formally establishing inter-firm networks, is represented by a purposeful, *functional* integration of value chain collaboration, as well as building up a competence network between collaborating firms. A distinction between clusters defined as either input-output relations or as networks is that *proximity* is the most important constituting variable in the first case, while networking represents a step towards more *systemic*, planned forms of co-operation. It also represents a shift from vertical to horizontal forms of co-operation, which more efficiently promotes learning and innovation in these systems. More systemic forms of co-operation can be developed by establishing systems, either in the form of production or innovation systems, which are characterised by *system* integration, which can itself, extend across time-space.

Types of regional innovation systems

It is important, analytically as well as politically, to distinguish between different types of RIS. Thus, Asheim (1998) distinguishes between three main groups of RIS in order to capture some conceptual variety and empirical richness in this phenomenon.

- The first type is *a territorially embedded regional innovation network.* The network covers firms and formal institutions in a region, the collaboration is based upon geographical, social and cultural proximity. The actors are embedded in local, socio-cultural structures facilitating close co-operation. The best examples of this kind of RIS (or more correctly regional cluster) is industrial districts Mark II, where firms base their innovation activity mainly on localised learning processes without much interference with formal institutions. This kind of network corresponds with the understanding of the interactive innovation model as knowledge flows both to and from the firms and includes several types of knowledge.

- Innovation networks may be further developed into *regionally networked innovation systems*. The firms and institutions are still embedded in a particular region and characterised by localised, interactive learning. However, the systems have a more planned and systemic character. The development from network to system occurs through the strengthening of the regional, institutional infrastructure; i.e. more R&D-institutions; vocational training organisations and other local organisation are involved in firms' innovation processes. The networked innovation system represents an attempt to increase innovation capacity and collaboration through public policy instruments. The networked system is more or less regarded as the ideal-typical RIS; a regional cluster of firms surrounded by a local 'supporting' institutional infrastructure.

- The third type of RIS, *regionalised national innovation systems*, is different from the two preceding types in several ways (see Figure 8.2). First, parts of the local industry and the institutional infrastructure are more functionally integrated in national or international innovation systems. Innovation activity to a larger extent takes place in co-operation with actors outside the region. A typical example may be regional clusters where the formal institutions stimulating firms' innovation activity are found mainly outside the region. The co-operation may, for example, take place between R&D departments in large corporations or advanced, smaller firms and national and international R&D institutes. Second, the collaboration is to a larger extent than in the two other types of RIS based on the linear model of innovation, as the co-operation mainly involves specific innovation projects to develop more radical innovations, with the use of scientific, formal knowledge. Co-operation may be stimulated when people have the same kind of education (e.g. as engineers), sharing the same knowledge and having common experiences and understandings, rather than belonging to the same local community.

The two last types of RIS constitute two main ways to further upgrade the innovation capability in regional clusters, i.e. to better connect firms and local network to regional, national and international knowledge institutions.

This conceptualisation of regional innovation systems is relevant to analyses of innovation support structures and policies. On the one hand, in principle, it informs analyses of innovative activity and the development of innovation policy instruments for a wide spectrum of firms and regions.

However, the question can be asked, "to what extent, if at all, ... [can] peripheral regions innovate?" (Morgan 1997, p.495). On the other hand, RIS is a useful theoretical construct for studying industrial development and industrial development strategies in particular regional clusters (Rosenfeld 1997). It may be less useful as an analytical framework for analysing and constructing development policies for rural and peripheral areas with little manufacturing industry and few traditions, and for declining industrial regions dominated by branch plant activities of TNCs. Regionalisation trends in these places appear to be structurally constrained (Asheim and Isaksen 1997, Pike and Tomaney 1999).

Main type of RIS	The location of knowledge institutions	Knowledge flow	Important stimulus of co-operation
Territorially embedded regional innovation network	Locally, however, few 'formal' knowledge institutions	Interactive	Geographical, social and cultural proximity
Regional networked innovation systems	Locally, a strengthening of (the co-operation with) knowledge institutions	Interactive	Geographical and social proximity as well as systemic university-industry co-operation
Regionalised national innovation systems	Mainly outside the region	More linear	Individuals with the same professional education and common experiences (organisational proximity)

Figure 8.2   **Some characteristics of three main types of regional innovation systems**

Some important criticisms of the research on industrial districts and flexible specialisation must be born in mind when considering the usefulness of the RIS concept (Amin and Robins 1990, Asheim 1992). There is a danger of generalising too broadly the potential of RIS on the basis on only a few empirical case studies. Regional innovation policies and instruments have emerged mainly from experience in successful places, like the Italian industrial districts and Baden-Württemberg. In addition, innovation is identified particularly in geographical clusters of small firms (OECD 1998), based on anecdotal evidence from these 'success stories'. It is possible that too much emphasis is being put on this experience in formulating general innovation policies, building regional technology support infrastructure, and making the development of RIS the preferred instrument of growth policy for all kinds of firms in all kinds of region.

The literature on RIS risks becoming the same 'pseudoconcrete analysis' which tends to "freeze, and then present as universal, relationships which are contingent and historically specific" (Sayer 1985, p.17). It needs to be emphasised that a learning-based strategy of endogenous regional development cannot be applied across the board. The necessary socio-cultural and socio-economic structures are found only in relatively well off 'motor' regions, and the techno-economic and political institutions are found only in relatively developed countries (Asheim 1998).

The second type of regional innovation system (RIS) underlines the importance for regional clusters – and in particular regional clusters of SMEs – to be "in touch, not necessarily directly, but through the supply chain, with global networks" (Cooke 1998, p.10). This is an objective which Camagni assigns to *innovation networks*, where "local firms may attract the complementary assets they need to proceed in the economic and technological race ... through formalised and selective linkages with the external world" (Camagni 1991, p. 4). The focus of the learning economy perspective is primarily on 'catching-up' learning (i.e. learning by doing and using) based on incremental and not radical innovation. Increasingly, what is needed in a competitive, globalised economy is the creation of new knowledge through searching, exploring and experimenting as well as through more systematic R&D for the development of new products and processes. However, especially in regional clusters dominated by traditional SMEs, relevant R&D competence may be scarce, and the dominant skill base (informal, experience-based knowledge) may represent a barrier to the adaptation and introduction of technical innovations (Varaldo and Ferrucci 1996). In particular, when firms in clusters face exogenous radical changes in markets or technology, R&D-competence is increasingly needed to produce radical innovative responses to external stress.

One way to improve the innovative capacity of regional clusters of SMEs is to stimulate co-operation with local R&D institutions, technology transfer institutions, and centres of real services, for example, which could systematically assist SMEs to keep pace with the latest technological developments. However, for SMEs to be innovative, there is often a need to supplement localised tacit and codified knowledge with the R&D competence and systematic basic research and development, typically undertaken in universities and research institutes. In the long run most firms cannot rely only on localised learning, but must also have access to more universal, codified knowledge of, for example, national innovation systems. This means that the strength of traditional, place-specific and often informal competence must be integrated with codified, more generally available, R&D-based knowledge. Thus, in order to attain more radical innovations, firms (and in particular traditional SMEs) in regional clusters may need strong and long-term relationships with external actors, so that firms have access to information and competence which may supplement local competence. Regional clusters can become locked-in to established technologies and dated products and solutions unless they receive impulses and competence from outside, especially through regionally networked innovation systems.

Empirical illustrations: three examples

Three examples of regional clusters in Norway are explored in this section to illustrate how firms exploit the geographical scale and scope of knowledge infrastructure and innovation systems to strengthen their competitive advantage. The clusters are Horten, Jæren and Sunnmøre (Figure 8.3). This section aims systematically to compare innovation performance in these three places, by using the same analytical framework of regional innovation systems in each case. The empirical illustrations are mainly based on interviews with enterprise managers, R&D managers and so on, undertaken in late 1998.[58] We target what has been seen by some observers (Maskell et al 1998) as a weak point in contemporary research on regional innovation systems, i.e. empirical validation of when and how regional resources facilitate learning and innovation.

Firms in different parts of mechanical engineering dominate the three regional clusters studied here. The branches of industry involved range from high technology (electronics in Horten) through medium-high technology (machinery industry in Jæren) to medium-low technology (shipbuilding in Sunnmøre) and low technology (the foundry industry at Jæren). The industries analysed contain a few larger enterprises (LSEs, with 250 employees or more), but are dominated by SMEs (Table 8.1). These enterprises also form local production systems, where some – and often large – end-use firms organise a production system including several local sub-contractors and suppliers.

Horten is a municipality of 24,000 inhabitants. The other regions comprise several municipalities, constituting local labour markets with 77,000 and 90,000 inhabitants. However, Jæren is also part of a functionally integrated labour market of 280,000 inhabitants. In Jæren, the enterprises studied all belong to the TESA network organisation (Technical

Co-operation), which includes the heart of the mechanical engineering sector in the region. In the other regions the analysis focuses on local sectors of specialisation, electronics in Horten and mechanical engineering in Jæren. The location quotients for these industries in these clusters are 10 and 5 respectively.[59] Horten and Jæren have about the same share of innovative enterprises as found in the electronics and mechanical engineering industries nation-wide. However, the shipbuilding industry at Sunnmøre has more innovative enterprises than this industry generally in Norway.

Table 8.1   Overview of the three regional clusters

	Horten	Jæren	Sunnmøre
Number of inhabitants in region	24,000	90,000	77,000
Industry studied	Electronics	Mechanical engineering	Shipbuilding and suppliers
Number of firms[60]	25	13	90
Number of employees	1,900	3,000	4,200
Firm structure	1-2 LSE*, the rest SMEs	4 LSEs, 9 SMEs	A few LSEs, most SMEs
Location quotient (ca)	10	5	8
Per cent innovative enterprises[61]	58	51	64
Per cent innovative enterprises in Norway	56	52	39
Number of firm interviews	11	9	7

* Large Scale Enterprise

Regional network: the Sunnmøre region

Sunnmøre is the largest shipbuilding region in Norway. At the beginning of 1999, however, several shipyards in the region experienced falling orders, leading to redundancies in other local firms. This was caused by, amongst other things, reduced activity and investments in the oil industry following a period of low oil prices, the traditionally lower levels of support for shipbuilding in Norway compared with its EU competitors, reinforced by relatively high labour costs in Norway (Hervik et al 1998). This shows how local industrial development is crucially dependent upon wider macro-economic conditions (such as oil price fluctuations) and policy decisions at the national and wider scales.

Apart from this recent decline, Sunnmøre has been among the winner regions for shipbuilding in Norway since the 1970s, experiencing job growth. The shipbuilding industry at Sunnmøre covered 4,200 jobs in 1997, of which 1,600 were found in 14 shipyards that concentrate on building different types of specialised ship, but a large number of temporary workers are not included in these numbers of jobs. The remaining 2,600 jobs were in 80 equipment supplier firms and ship designers.

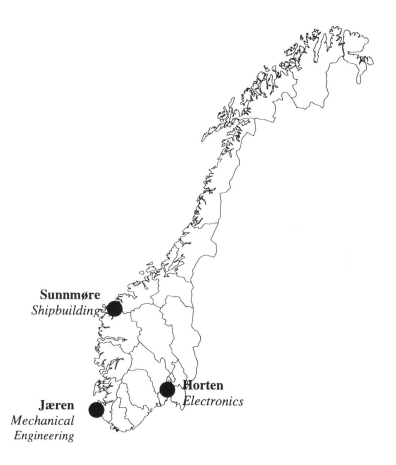

Sunnmøre
Shipbuilding

Horten
Electronics

Jæren
Mechanical
Engineering

Figure 8.3   **Location of the three study areas**

The competitiveness of the shipbuilding industry at Sunnmøre, revealed in the firms' high export share, is to a large extent based on the innovation capability in the cluster (Table 8.1). Incremental innovation activities and learning occur on the 'practical' level, which includes new and smart solutions in the construction of products and in the 'daily' work of creative and skilled workers and engineers, leading to step-by-step improvements of existing products and production methods. In several cases, every new order implies a certain degree of innovation, where firms must improve existing products.

Incremental innovation occurs in four ways in this industry:

- through local user-producer interaction;
- through incremental innovations on the shop floor;
- through local knowledge spill-over; and
- through means of co-operation with local organisations.

User-producer relations

A main driving force behind the continuos incremental improvement of products is to satisfy the new demands and needs of customers and users. Thus, for equipment suppliers, local shipyards are important sources of innovation. Local shipbuilding consultants, who design and construct new ships, also have an important role in mediating the demands and specifications of products between yards and equipment suppliers. Shipyards have also co-operated in the long-term with ship owners who often return to the yards to discuss new solutions and build new ships.

In the shipbuilding industry, we can also distinguish between customers and users as sources of innovation. The 'end' customers are the shipping companies, while individual users are fishermen and seamen. Discussions with skippers, chief engineers and other crewmembers give important feedback on how the firm's (and its competitors) products work, generating suggestions for improvements. Sunnmøre is an important fishing district, especially for fishing in remote waters, which place tough demands on ships and equipment. It is also an important shipping area, especially for ships serving the offshore activity of the petroleum industry. The contact with users occurs when fitters, service workers and product developers visit shipping companies or working ships. However, a lot of contact takes place when people meet in their spare time, or meet on ferries and in airports, and then discuss how different products work, what to do better. The contact is facilitated by, for example, seamen and product developers being part of the same local culture and sharing some common knowledge and experience. The producers are aware of signals from the local users, and use that experience in the step-by-step improvement of products. Some producers have formal procedures to collect, assess and proceed with ideas from customers and users. The feedback loops from the users and customers are thus of vital importance for the incremental improvement of products.

Incremental innovation

A second main way in which innovation occurs at Sunnmøre is as *incremental improvements on the shop floor*, relying on the experience-based competence of engineers and workers who know the firms' products and technology well. This kind of innovation also reflects a common interest in the local community in developing the shipbuilding industry and the civil society. This attitude is seen in the drive, enthusiasm and loyalty of the work force, i.e. when workers exert themselves to find better ways to do things, leading to frequent, small innovations. These are attitudes rooted in the way the shipbuilding industry was established and developed in the area. Local entrepreneurs mainly started the firms to supply a local fisheries market and, later on, oil exploitation in the North Sea. Several firms are spin-offs from existing firms and especially from the shipyards. Thus, firms have grown out of the local milieu, supporting a local market, employing local competencies, and using local capital from community established banks, relatives and, in some cases, also from workers. The workers considered themselves as a proletariat only to a small degree, and the organised labour movement (that has been and still is very strong in Norway) has traditionally attained only a weak foothold in this part of the country. Entrepreneurs, firm leaders and workers share the same attitudes built on self employment and stimulated by traditions of collective entrepreneurship through co-operatives (Wicken 1994).

Incremental innovations on the shop floor may, however, be challenged by the globalisation processes, illustrated by the British corporation Vickers that bought the region's largest shipbuilding and maritime firm, Ulstein, at the end of 1998.[62] The loyalty and enthusiasm of the workforce are partly determined by the local character of the firm, as an integrated part of the local community. This solidarity with the workplace may disappear when firms are now longer owned by local entrepreneurs, thereby damaging the competitiveness of the firm in the long run. A high labour turnover rate may also hamper incremental innovation on the shop floor. Thus, late 1990s (until 1999) have seen a shortage of qualified labour which has resulted in the high labour turnover (up to 28 percent in 1997 in one of the largest shipyards) and firms hiring workers on such a temporary basis. In such a situation it may be more difficult to build up loyalty and experience among labourers as a basis for innovation on the shop floor.

Knowledge spillover

Knowledge spillover and the exchange of new ideas between local firms further stimulates innovation and learning. The cluster contains a varied set of specialised firms, implying that firms may find other local firms to consult and/or buy specialised competence from. The area also holds several innovative firms. New ideas may be quickly copied and further developed by other local firms. Knowledge spillover takes place when firms co-operate on specific project, when firms obtain advice from neighbouring firms, in personal contacts between workers in different firms, and through job shifts. Experience-based competence is transferred in informal circumstances outside working hours. The Sunnmøre area contains firms with similar products and technologies. Employees meet privately, and may then discuss good and bad experiences with the use of a particular machine, how to solve specific problems, and so on. A lot of information and new ideas are captured in such circumstances, and are than tested and further developed in other firms.

Local institutions

The fourth way in which innovation occurs is by means of *co-operation via public/private institutions*. Thus, a broad knowledge infrastructure of vocational schools, a technical college and three associations, *Verkstedforeninga i Ulstein distrikt* (The Mechanical Engineering Association in the district of Ulstein), *Maritim Nordvest* and *Nordvest Forum*, stimulates local co-operation, competence building and some innovation activity in production methods. The Associations are established by and for local firms. The workings of these organisations mean that the shipbuilding industry at Sunnmøre can be viewed as a regionally networked innovation system. However, local firms rarely use local R&D-institutions or higher education institutions when innovating. As a consequence the cluster is not an ideal-typical regional innovation system. To be one would involve a much greater local innovative collaboration between local industry and the R&D milieu.

    The example of the Sunnmøre region has so far focused on incremental innovations stimulated by local co-operation, knowledge spillover and the existence of experience-based knowledge. However, innovations increasingly involve *the use of R&D-based knowledge*. Thus, several firms put more effort into research and development, to go beyond mere incremental innovation activities in order to fulfil customers' demands. Several larger firms have established R&D departments to accomplish more basic product development, for example to increase the speed and reduce the weight of ships and to use new materials. This kind of innovation activity is not triggered by orders from specific customers. The firms want to be ahead of demand, often discussing new concepts with potential customers at an early stage.

This innovative activity is organised in R&D departments, but takes place in co-operation with other sections of the firm, especially the engineering and marketing departments. Firms also co-operate with external R&D-institutions, most often with SINTEF in Trondheim, the largest technical research institution in Norway. Some firms collaborate with similar institutions in other countries. Ulstein Propeller (producing propellers and propulsion systems, with 380 employees at Sunnmøre) participates in a large research project in the US, in co-operation with MIT and the University of Texas, the US Navy and other American engineering firms. The large and/or advanced firms in Sunnmøre compete on international markets, and have to co-operate with the most competent R&D milieus in their speciality. Thus, firms are integrated with national and even international innovation systems. Several firms have taken out patents, including small firms, such as Jets Vacuum with 25 employees (producing vacuum pumps) and Fiskerstrand Verft with 115 employees (producing wave damping systems, developed in co-operation with SINTEF).

Firms make use of both local competencies and internationally available knowledge and skills. To be able to continue as a leading producer of advanced ships, the second largest shipyard in the region, Kværner Kleven, see it of utmost importance to be located in a maritime milieu and using the full range of shipbuilding competencies available in this milieu. The milieu, as they understand it, includes both the regional shipbuilding cluster, with users, suppliers, and a competent work force, as well as access to the competence of and co-operation with SINTEF and other parts of Kværner International.[63]

Technological Co-operation: The Case of Jæren, Norway

The second regional cluster is located at Jæren, which is a region south of Stavanger, in the county of Rogaland, in the southwestern part of Norway (see Figure 8.1). Rogaland is the largest industrial county of Norway, dominated by the oil and gas industry and the mechanical engineering industry, traditionally specialising in the production of farm-machinery. It is the latter sector that constitutes the cluster under study. This industrial cluster, which for many years has been very competitive and export oriented, has undergone considerable changes during the last ten years due to the globalisation of the world economy. ABB's acquisition of Trallfa Robot in 1988, now called ABB Flexible Automation, is Europe's leading producer of painting robots for the car industry. This was the first major example of FDI in the region, while Kverneland, the world's largest producer of agricultural equipment is the main example of a local firm becoming a TNC. Kverneland today has production facilities in 14 countries, and has during the 1990s bought factories in Italy, Denmark, Germany, and the Netherlands. Other examples of FDI in the region are the Swedish Grimaldis' Cycleurope take-over of Øglænd in 1995[64] and the British Williams Group buy-up of Norsk Hammerverk in 1998.

Jæren is an area of specialised production with a high degree of inter-firm networking and co-operation. This co-operation is institutionalised through TESA (technical co-operation), which was established by local firms in 1957 as an organisation with the aim of promoting technological development among member firms, which were small and medium sized, export-oriented firms producing mainly farm-machinery. This has, among other things, resulted in the district today being the centre of industrial robot technology in Norway with skills in industrial electronics and microelectronics far above the general level in Norway. Furthermore, the use of industrial robots is much more widespread in this region than in the rest of Norway. It has approximately one third of all industrial robots in Norway but with only 3 percent of Norway's industrial employment.

In the Jæren region, the total turnover for the network is about 3 billion NOK with more than 3,000 employees. These numbers increase substantially when affiliates and subsidiaries outside Rogaland are included. The network involves companies operating very different industries - the production of farm-machinery and equipment (Kverneland Klepp and Kverneland Underhaug), advanced electronic equipment for paramedics (Laerdal Medical), production of bicycles (Øglænd), industrial (painting) robots (ABB Flexible Automation), china (Figgjo), advanced naval electronics (Simrad Robertson), advanced hydraulics (Bryne Mekanikk), and fire-equipment (Norsk Hammerverk). The TESA firms are strongly export orientated and on average export 70 percent of their output. However, some firms export more. Three firms exported 90 percent of their output in 1997 (Laerdal 95 percent, ABB Flexible Automation 88 percent, and Kverneland 93 percent). According to the firms, without the inter-firm technological co-operation taking place within TESA, the development of this very strong competitive advantage would not have been possible.

The TESA firms vary in size from 800 employees in the largest manufacturing units (i.e. Kverneland Klepp) to 50 employees in the smallest (Norsk Hammerverk). Sectorally, the TESA network is still dominated by manufacturing industries. However, the network is heterogeneous within this broad sector ranging from the production of fire equipment and bicycles to industrial robots and medical equipment, and the more traditional production of agricultural equipment. If the production of the firms is classified according to R&D intensity, following the OECD typology of high tech, medium-high tech, only high tech production is missing.

To promote member firms' competitive advantage, TESA took an active part in the establishment of JÆRTEK (Jæren's technology centre) in 1987. The aim of JÆRTEK is to offer training to prepare workers in firms, as well as pupils in technical schools, for the advanced industrial work of today and tomorrow, and to secure the competence base for continued, rapid technological development. To achieve this, the first complete computer-integrated manufacturing (CIM) equipment in Norway was installed in JÆRTEK.

The close, horizontal inter-firm co-operation in this district with interactive learning, resulting in the development of core technologies is unique in an international context. Technological co-operation was strongly dependent on high levels of internal resources and competence in the firms, but did also involved regional and national R&D institutions (e.g. Rogaland Reseach in Stavanger, Christian Michelsens Institute in Bergen, and The Technical University and SINTEF (Center for Industrial and Technical Research) in Trondheim) in the development of the robot technology.

Today, the focus on technological co-operation has been somewhat reduced, and the focus on organisational, managerial and strategic issues has increased. The issues discussed in the network relate to projects for developing competencies, the use of information technology in the network, further development of the co-operation model of the network and new approaches from TESA for it to remain as a driving force in the regional innovation system. Thus, today TESA's main activities are; a) to implement and organise common development projects; b) organise subject groups in productivity, purchasing, quality etc; c) to act as a forum for Managing Directors; and, d) to provide for leader training.

This situation is the result of two parallel trends in globalisation, the formation of local TNCs and the new local presence of foreign TNCs. Originally, all of the TESA firms were established as local initiatives. During the last 10-15 years some have been bought up and transformed in to subsidiaries of TNCs while others have grown to be multinational companies themselves. Examples of the first category are ABB Flexible

Automation and Øglænd DBS, while Laerdal Medical and the Kverneland group are examples of the second category. All the companies in TESA have been greatly affected by the constant drive towards globalisation and their transformation into TNCs controlled by external forces with minimal focus on regional or local issues.

This development has resulted in many of the firms growing out of the district, in terms of technological development. The mechanical engineering firms at Jæren are more limited in their co-operation with national R&D institutions because they often do not have the competence the firms need. The national innovation system in Norway largely supports Fordist process industries as well as the oil and gas industry that has resulted in a typical linear innovation system. Consequently, the national innovation system has neither the competencies that the mechanical engineering industry needs nor the interactive processes that normally generate innovation in the industry. In this situation, most of the firms in Jæren use two strategies to find relevant R&D-based competencies. The first strategy is to use foreign innovation systems, especially specialised R&D institutions and universities in Sweden and Germany, where manufacturing industry is large. The second strategy is to utilise R&D departments within the corporations to which they belong or co-operative research with foreign, strategic partners.

The merging of firms into larger corporation is a challenge for the TESA network as an arena for local inter-firm co-operation. As the member companies become less independent and reorientate towards their transnational corporate owners, centrifugal forces in the network become stronger. In addition, through their acquisitions new external transnational owners "can enter the web of relations between firms within the districts and integrate them into their corporate network" (Tolomelli 1990, p. 366).

To counter this, another rationale for building strong regional development coalitions, is to increase the importance of local, innovation networks, so that member firms can compete not only with competitors outside the region, but also with other members of the corporations of which they belong. Global competition and the possibilities and threats of relocation of production plants represent a constant challenge. Policies promoting the learning capacity and innovative capabilities of localized networks are thus essential for local firms to be competitive.

The TESA office is now located in the new science park in Stavanger, close to Rogaland Research and the Stavanger Regional College. This will strengthen the close relationships between both research institutes, other centres of competence, local public authorities and educational institutions. TESA could still have a potentially important role to play in the future in promoting the industrial renewal necessary to upgrade some of the more traditional firms in, for example, the farm-machinery industry to higher value-added production. However, as a result of globalisation processes more and more of this work has been internalised in the corporations, and there is a tendency for innovation systems abroad to substitute for the regional and national ones. Nevertheless, the local production system (e.g. suppliers and sub-contractors) and the local labour market still seem to present important competitive advantages. Thus, an important question for TESA concerns the future role of local knowledge and localised learning processes in local upgrading processes in the face of globalisation and the spread of TNCs. The future development in the region will depend on this local upgrading of competitive advantage among local firms in general (in which TESA can play a part), and specifically on the behaviour of the TNCs investing in the region.

The electronics industry in Horten – commercialisation of 'national' research

The electronics industry in Horten comprises some 1,900 jobs and 25 firms at the end of the 1990s, making it one of the largest electronics clusters in Norway. The motive power force in the cluster comes from large system houses and OEM suppliers (original equipment manufacturers). The system houses control their own, highly advanced products, often produced in small batches, and sold to final customers on national and international markets. Currently, Horten has nine system houses with a total of nearly 1,000 employees. The largest (Kongsberg Maritime) employs 700 people. The OEM suppliers have their own products as well, but these are components used by mostly international system houses. Horten has three OEM-suppliers employing nearly 450 people. In addition, the local industry includes 13 sub-contractors with 500 jobs, serving system houses and OEM suppliers in Horten, other parts of Norway and Scandinavia. The sub-contractors include contract suppliers of electronics components as well as mechanical and electro-mechanical firms.

The systems houses and OEM-suppliers are very much focused on *product innovation*. Many firms in Horten develop patented products, new to their niches on the world market, and the firms generally have relatively high R&D costs. Most of the large firms are world leaders in their niches.

The innovation activity in the system houses and OEM suppliers mainly takes place in co-operation with national and international R&D-institutions. Large, mainly national customers also form part of the innovation system because they act as early and demanding customers through testing prototypes, giving feedback and making claims on products. Other advanced firms and suppliers complement the firms' internal competencies in innovation projects. Still, much of the innovative activity and learning takes place inside firms with 'large' R&D-departments and 'many' engineers, and in co-operation with other units in the corporations the firms belong to. The firms hold equally high competencies in specific niches as do national R&D institutions. Thus, some researchers in the firms publish scholarly articles and present papers in international conferences based on development projects in the firms.

Commercialisation of R&D-results

The system houses and OEM suppliers in Horten are part of a national, technological innovation system. Firms were originally established through the commercialisation of R&D results from Norwegian technological R&D institutions, and have a rich history of interaction with these institutions. One of these institutions is situated in Horten itself (a branch of the Defence Research Institute), the others are in Oslo and Trondheim. Individuals, competencies and property rights over knowledge have been transferred from these institutions to private firms, either new firms or established firms. Today, much more interactive learning occurs, through co-operation between firms and R&D institutions (that generally are co-funded by the Norwegian Research Council and increasingly by EC) on specific projects, and through the mobility of individuals between the firms and the R&D institutions. The flow of individuals and knowledge occur in joint innovation project, along with job shifts, recruitment of new candidates, upgrading of employees in the firms, and by highly qualified researcher in the firms supervising doctoral students in R&D institutions and universities. Recruitment of researcher from the R&D institutions facilitates continued collaboration between firms and the institutions through establishing personal relations.

Initiatives by public authorities have been important in stimulating the innovativeness and competitiveness of the electronics industry in Horten. An explicit national effort to create a knowledge-based Norwegian electronics industry lies behind much of the development taking place in Horten. The vision to create a high-tech Norway, and the national instruments to accomplish the task, were long based on a linear view of innovation: advanced research in national R&D institutes should result in new technology, new products and new firms. This technology push has characterised product development in the firms in Horten. It is based on new product ideas in firms' R&D departments and their vision of creating products to meet demand.

The electronics industry in Horten constitutes a *regionalised national innovation system*. However, systems firms and OEM suppliers have partly grown out of the national system that they rose from. This has been brought about through collaborating on product development with foreign R&D institutes, firms and other units in the corporations they belong to. Vingmed Sound is, for example, the only European producer of ultrasound products, and SensoNor is the world's largest independent producer of collision sensors for the automobile industry. Research institutes quire both firms to be industrial partners in EC-financed projects. The recruitment of researchers in several firms also increasingly takes place at the international level, as there is a shortage of research engineers in Norway.

Foreign corporations own three of the system firms and OEM-suppliers in Horten. The firms, however, reign supreme in their core technology and competence inside the corporations. For example, AME Space is the only firm inside Alcatel to master the SAW technology (Surface Acustic Wave). The competence is embedded in human capital and in personal relations between researchers in the firm and in Norwegian R&D institutes. Thus, the competence may be difficult to transfer from Norway.

Increased significance of the local level

Business relations in Horten are historically embedded in national, and increasingly in international, rather than local social structures. Nevertheless, the local/regional level has been of increasing significance for some aspects of innovation activity in the electronics industry in Horten since the 1980s. The importance of the local level centres on the unique competence in the workforce of this area, gained through years of trial and error in innovation projects. Some firms collaborate with the public sector local employment office in the organisation ARENA, where workers temporarily released from one firm are being soaked up on short-term contracts in other firms, instead of possibly leaving Horten. Local sub-contracting started in 1980 as a result of system firms closing down most of their in-house production. In some cases, managers of the production departments in the system firms have established their own firms, sub-contracting to their 'mother' firm. These generally local sub-contractors now carry out most of the physical production. The system houses focus on assembly and final testing of the products, as well as product innovation. The local sub-contractors are not involved in product innovation as such. However, they are increasingly important in industrialisation, i.e. in the task of transforming prototypes into effective industrial production. The sub-contractors have been involved in industrialisation at earlier phases the last ten years. Instead of just receiving drawings and documentation from their customers in order to produce, the sub-contractors increasingly give advice and comment on drawing and design before the product is finally developed. The intention is to obtain products that can be produced and tested effectively, using the cheapest usable components. The sub-contractors advice is based on their skilled workers and engineers competence and experience in production technique and their knowledge of components and materials. Organisational innovations occur in the local production system through more long-term and binding co-operation between system firms and suppliers, and more 'just-in-time' deliveries to some system firms. Location close to suppliers is an advantage to industrial development as well as to start ups of new production processes, although this is not a matter of necessity. However, it is easier to organise fast and frequent meetings to discuss solutions and undertake changes in a new production start-up, if important suppliers of components and modules are close by. Thus, the electronics firms in Horten have become more embedded in a local industrial milieu through closer interaction with local sub-contractors.

Conclusion

We began this chapter by referring to the two development tendencies of globalisation and regionalisation; the functional integration of firms in global networks, and the simultaneous reliance on place-specific factors and regional collaborative networks in creating competitive advantage. From this we proposed a multilevel approach to innovation systems to analyse regional economic clusters.

The three case studies underline the relevance of a multilevel approach as firms exploit *both* place-specific local resources and institutions as well as external, world-class knowledge to strengthen their competitiveness. However, there were differences between the three clusters in how firms came to make use of global, national and regional/local resources in innovation processes, reflecting local specialisation.

The national level has traditionally been of greatest importance for the competitiveness of the electronics industry in Horten. Most of the system houses and OEM suppliers were established through the commercialisation of R&D from Norwegian technological institutions. Horten was the place where the pioneer firms came to be located, while their important R&D contacts and customers were in other parts of

Norway. The global level has become more important as firms have been bought up by TNCs reorienting co-operation towards other firms and R&D departments inside these corporations. Some firms also form strategic alliances with foreign firms or collaborate with foreign R&D institutes in EC-funded projects. The local level has become more important through the formation of a specialised local labour market, out-sourcing by systems firms and the formation of closer collaboration between systems firms and sub-contractors. The systems firms and OEM suppliers still rely heavily on the national innovation system for product development, while they increasingly consult local sub-contractors on manufacturing processes.

At Jæren and Sunnmøre the local and regional levels historically have been decisive for technological development and competitiveness. The mechanical engineering and shipbuilding industries in these areas were established to serve the local agriculture and fishery sectors. Firms have developed competitive products to meet the demand of local farmers, fishermen and ship owners (i.e. through local user-producer interaction). In both clusters, local organisations have stimulated local collaboration and have promoted the formation of shared, locally specific competencies of both a tacit and codified nature. Both areas are also regarded as entrepreneurial districts, where a predisposition to self-employment encourages people to start their own businesses after gaining experience in other local firms.

Many firms in Sunnmøre and in particular in Jæren have now grown out of those districts, in terms of technological development. This has occurred as a result of globalisation and the development of transnational corporations in the last ten years and the limitation of the regional R&D system. As some firms in Jæren and Sunnmøre are world leaders in their niches, necessarily they co-operate with the 'best' R&D milieus, which they find at the national and international levels.

The use of external competence networks in all the three clusters demonstrates the importance of national innovation systems, the existence of world leading national research groups and collaboration with global actors, for innovation processes in regional clusters. The extensive collaboration with national and international actors reveals that ideal-typical regional innovation systems, i.e. regional clusters 'surrounded' by supporting local organisations, are rather uncommon. At least, this seems to be the case in a country like Norway, with comparatively small regional industrial milieus but large, national R&D-institutions and low cultural barriers between those national institutions and industry. Accessing external resources outside the local community is crucial for many SMEs. Thus, it may be necessary to moderate the assertion that small firms are very dependent upon the local industrial milieus in promoting innovation activity.

It is, nevertheless, vital to underline innovative activity as a regional phenomenon in the kind of clusters studied here. Especially in Sunnmøre and Jæren, regional resources and collaborative networks have large, and in some cases decisive, significance for firms' innovative activity, i.e. the regional level is important even if the clusters do not constitute ideal-typical regional innovation systems. What local resources are important differs between regions. However, those resources include unique combinations of the knowledge and skills of the labour force and specialist suppliers, the existence of local learning processes and spillover effects supported by geographical and cultural proximity, as well as formal co-operative institutions.

The three case studies also stress the importance of localised knowledge, including formal knowledge. Formal, scientific knowledge is vital in the kind of product development carried out in, for example, the electronics industry in Horten. However, the experience-based knowledge of key people, as well as artisan skills, supplements that scientific knowledge. Informal knowledge is gained through years of trial and error in product development and in personal networks within R&D milieus, customers, users and so on. This kind of knowledge comes without full documentation. The combination of these different kinds of knowledge is bound up with individuals and cannot be moved without people also moving. This place-specific knowledge includes tacit knowledge as well as disembodied codified knowledge. The interactive way, in which this knowledge is acquired, is an important explanation of the tendencies of successful path dependence observed in several regional clusters. These factors in combine produce a context of untraded interdependencies that enhance the competitiveness of firms.

9 Enterprise, Power and Embeddedness: An Empirical Exploration

Michael Taylor

Introduction

The purpose of this chapter is to explore the impact of power inequalities between firms on the nature of their interrelationships and especially on the nature and form of their transaction structures. Power itself is a slippery concept, frequently invoked as explanation but rarely dissected and explored. It is recognised only indirectly in the industrial districts literature (see Grabher 1993a), where greater emphasis is placed on trust, reciprocity, loyalty, collaboration and co-operation as mechanisms of growth. Indeed, the impression is created that symmetrical rather than asymmetrical relationships between firms are the foundations of local growth in an era of global economic relationships.

It is contended in this chapter that unequal power relations and power asymmetries are, however, fundamental to the nature and functioning of inter- and intra-organisational relationships. They circumscribe action and individual agency. They mould and modify information exchange and, therefore, impact upon learning processes and innovative capacities. These relationships are also constituted within complex and diverse networks which, as Dicken and Thrift (1992, p. 279) have maintained, reflect the technical, organisational and geographical characteristics and dimensions of production chains. Power asymmetries engender differential growth among firms within these networks. They create winners and losers and generating change. It is these relationships that need more fully to be incorporated into the 'industrial districts' literature, and into work on local social capital formation as a foundation of local growth (Maskell et al 1998).

The argument of this chapter is developed in five stages. First, the evolving notion of business enterprise power networks is developed and contrasted with the inter-firm relationships that lie at the heart of the 'industrial districts' literature. Second, the concept of power is explored in the context of business enterprise relationships. Third, an attempt is made to operationalise these perspectives on power in the empirical context of 90 Australian engineering companies, to identify inter- and intra-organisational dimensions of control and autonomy through multivariate analysis. Fourth, the relationships between the revealed dimensions of power and authority and the transaction structures of the sampled business enterprises are examined. Finally, the significance of these findings for the 'industrial districts' and 'learning regions' interpretations of regional economic growth is discussed.

Networks, embeddedness and 'learning'

In the rapidly expanding literature on industrial districts, networks, 'learning regions' and innovative milieu, it is widely contended that in a new era of flexibility (Piore and Sabel 1984, Sabel 1989) local economic growth is dependent on the incorporation of firms into socially embedded networks of collaborative production (Cooke 1998). Those networks are, in turn, "buttressed by a supportive tissue of local institutions" (Powell and Smith-Doerr 1994, p.370). The benefits of these collaborative and co-operative relationships are seen in terms of heightened place-based capacities for learning, information exchange, technological change and innovation amongst network members. Industrial districts are, by this interpretation, regional innovation systems, as discussed in the previous chapter, are engines of economic growth and competitive strength fuelled by intensified and localised processes of Schumpeterian creative destruction. They are dynamic nodes of innovation and learning incorporated into an emerging global mosaic of regions (Scott and Storper 1992).

Fundamental to the collaborative manufacturing of industrial districts and new industrial spaces are the processes of enterprise 'embedding', with Powell (1990, p.300) contending that all economic exchange, "... is embedded in a particular social structural context". By this view, anonymous markets are non-existent and economic life and transactions are rife with social connections (Granovetter 1985).

At the heart of the embeddeness thesis is *structural embeddedness*, which identifies the manner in which business enterprise relationships are articulated one with another and incorporated into networks. Structural embeddedness has four essential characteristics; *reciprocity, interdependence, loose couplings* and *asymmetric power relations* (Grabher 1993a, p.8-12). However, in much of the most recent literature on industrial districts and learning regions, the impact of asymmetric power relations on network relationships has been all but ignored. Certainly, co-operation and collaborate in industrial districts have been recognised but not the more brutal exercise of power, domination and subordination. The current approach to understanding the dynamics of industrial districts, devoid of power asymmetries, has been succinctly summarised by Brusco (1986):

> "[T]hree factors contribute to the success of an industrial district. First, there must be *competition* but also *co-operation*, which is a repeated game, based on a system of incentives. Second, there must be some level of *conflict*, to prevent paternalism, as well as *participation* to share experience and to ensure that the workers' capacity for invention is available to the firm. Third, two different kinds of knowledge, the *local and practical* and that of *science* must be connected. Local knowledge must be valued as even small improvements may open huge markets. *Institutions* must seek to support and promote these lessons." (Brusco 1996, p.118, emphasis added)

Thus, co-operation must be tempered with competition. Some activities in an industrial district might best be undertaken collaboratively and co-operatively while others might be best left to price determined competition (Cooke 1998, Enright 1996).

What this model of industrial districts offers is a vision of local competitive strength being created by the interplay of collaboration and competition (but just enough), labour market participation and conflict (but just enough), a mixture of tacit and scientific knowledge (but just enough) and institutional support (but just enough). How much is 'enough' is not specified in the model beyond the vagueness of being 'contingent'.

In the model, domination and control are pushed into the background. The model seriously underplays differences in the resource bases that firms accumulate over time and the nature of established relationships. It also neglects the evolving patterns of domination in the transaction structures of production chains.

Inequalities and power asymmetries between business enterprises have, in fact, been interpreted as undermining the collaborative structure of industrial districts and regional innovation systems. Herrigel (1993) has identified the local embedding of parts of large corporations in industrial districts in Germany as a mechanism through which those firms can retain power in a decentralised world order (p.247). Grabher (1993a) has maintained that large corporations who decentralise their internal structure to gain the strengths of an industrial district will, through those actions, open a district "to and insidious erosion of their specific supportive tissue of social practices and institutions" (p.24). Amin (1993) has gone further and has argued persuasively that networked localities, "are not the masters of their own destiny in being subject to the authority and influence of the prime-movers [global oligarchies dominated by TNCs] within the networks" (p.291). Markusen (1996) has interpreted the dynamics of Silicon Valley in a similar way.

Currently, there is only anecdotal information on how the complex relationships of dependency, control and power impact on industrial network relationships (Håkansson and Johanson 1993, p. 48). The analyses reported in this chapter are an initial and exploratory attempt to begin such a process of 'unpacking' by examining empirically the impact of power on the transaction structures of Australian engineering firms. To undertake the analyses proposed here involves a number of preparatory stages. First, the concept of power needs to be unpacked, and here interpretations of power as 'commodity', 'relationships' and 'dominance' are particularly valuable (Clegg 1989). Second, these notions of power need to be incorporated into conceptualisations of network structures, and in this context the analysis of structure produced by Dicken and Thrift (1992), and in Taylor and Thrift's (1982a, b, 1983) elaboration of enterprise segmentation, offer useful starting points. Third, a methodology must be devised to capture through measurement these ideas on network power. Here, contingency views on organisational structure, and the pioneering work on power of the Aston Group in management science is of particular relevance. From these foundations it is possible to build an exploratory statistical analysis of the impact of asymmetric power relationships on business enterprises' transaction structures.

Power and the business enterprise

Power is a slippery concept with multiple layers of meaning that are difficult to pin down with precision. A starting point from which to begin to tease these layers apart to better understand the nature of power and its asymmetries is the various ways in which power has been defined. Power has been conceptualised in a wide variety of different and quite distinctive ways in economic sociology. Three of those perspectives are helpful in coming to grips with the nature of the asymmetry of relationships that bind business enterprises into networks (see Allen 1997, Clegg 1989, Taylor 1995,1996).

The first and dominant view of power is of *power as agency*, a Hobbesian interpretation in which power is a commodity which business enterprises can acquire and use. Power, then, is a property of resources and the control of those resources by the enterprises that are dependent upon them (Pfeffer and Salancik 1978, Pfeffer 1981). These resources can include funds, personnel, information, products and services (Aldrich 1972), though in essence they all reduce to 'money' and 'authority', which can be labelled respectively as technical power and positional power (Benson 1975). By this interpretation, a firm has power to the extent of the resources it controls, with the additional condition that "... A has power over B to the extent that he can get B to do something that B would not otherwise do" (Dahl 1957, p.203). For Wrong (1995) power under these circumstances must also involve intentionality though this is a mute point.

Power, by this definition highlights, resource-based inequalities, working practices and structures of dominance that are both time-specific and place-specific. These are the standing conditions - normal business practices - in a place that are negotiated or imposed but nevertheless tolerated for longer or shorter periods to provide businesses, entrepreneurs and decision-makers with the certainty they need before they will commit themselves to production, consumption and investment.

A second and complementary view of power is of *power as relationships*. This is a contingent, Machiavellian interpretation of power reflecting what power does rather than what power is. It is power as influence, providing legitimation for some, granting status and bestowing membership on agents as individuals and as business enterprises. It can be seen as the power of information and rules. Fostering these rules of meaning and membership is regulation - the formal, codified regulation of law and 'real regulation' (Clark 1992, Marsden 1992) and the equally important but informal codification of local modes of social regulation (Jessop 1990). Here, according to Clegg (1989) is the power to create conformity, stability and isomorphism and to create styles of organisational structure and management conduct, and possibly differential performance, amongst sets of networked business enterprises.

Finally, there is the conceptualisation of *power as discipline*; the dominance and discipline of the state, society, capital and culture. Here is the empowerment and disempowerment of business enterprises as techniques of production and discipline in the workplace change through technical and managerial innovation and the shifting of those new techniques and practices from one place to another through processes of internationalisation and globalisation (Clegg 1989, Taylor 1995). In this Foucaudian interpretation of power, technology may radically alter control of the labour process or factory regimes, or TNCs may bring new work practices and transaction structures to a place, disrupting or even destroying past certainties. This view of power identifies the discipline and dominance of knowledge as it impacts on business enterprises. Owing to their own path-dependence, those enterprises are limited in the responses they can make by their accumulated baggage of past information and decisions, past experience and past performance - their sunk costs most broadly defined. Power as discipline, therefore, has the potential to introduce revolutionary change into economic systems at all scales

Powell and Smith-Doerr (1994) offer a complementary but individualised summary of these three forms of power as they might relate to dyadic relationships within enterprise networks, labelling them as formal authority, informal influence and overt domination:

> "Authoritative power involves issuing orders and instructions, with the expectation of uncontested compliance ... The source of the orders rather than their specific content induces compliance. Influence involves transmitting information from one person to another that alters the actions the latter would have pursued in the absence of information ... and domination entails the control of the behavior of one individual by another who can offer or restrict benefit or inflict punishment." (p.376)

For Clegg (1989), these three forms and layers of power are linked and interwoven as dynamic circuits, simultaneously facilitating and forming, cementing and stabilising, while at the same time disrupting and destroying economic and social relationships (see Taylor 1995).

Networks and power

As the discussion of the previous section has implied, however, power is fundamentally contextual. Individual firms do not exist in some impersonal ether but are, as Dicken and Thrift (1992) have argued, arranged within production systems and along value added chains through which they are both co-ordinated and controlled. These chains are complex sets of networks of interrelationships composed of relations and ties that have both content (information, advice, personnel and all kinds of resource) and form (strength). Co-ordination and control is, in turn, organised from centres of strategic decision making - principally large corporations whose influence extends beyond their own legal boundaries and permeates the operations and strategies of other network members (Cowling and Sugden 1987). Indeed, Powell and Smith-Doerr (1994) have pointed out that extensive research now documents how a firm's location within an inter-organisational network, frames and accounts for its structure and strategy.

This is not to suggest that networks share a common configuration. They can be very varied and, at the very least, inter-organisational networks have been categorised as 'centralised', 'balkanised' or 'disorganised' (Powell and Smith Doerr 1994, p.377).

Importantly for the present analysis, all three forms of power can be recognised in enterprise network structures related to the position of business enterprises and parts of business enterprises within them. There is also strong evidence to suggest that the distribution of power and the nature of relationships within networks may frequently remain stable. An inference to be drawn from this is that, "the basic units in a system of power are not individuals per se, but the statuses occupied by them and the relations and connections among their positions" (Powell and Smith-Doerr 1994, p.377). Thus, "[a] position's power - its ability to produce intended effects on the attitudes and behaviors of other actors - emerges from its prominence in networks where valued information and scarce resources are transferred from one actor to another" (Knoke 1990, p.9).

This is also the view of network power relations implied in Taylor and Thrift (1982a, b, 1983) in the business enterprise segmentation model they elaborated in economic geography. The foundation of this enterprise segmentation model is a typology of enterprise asymmetries that reflects the nature of unequal power relations based on:

- access to and control of resources of all types (power as agency);
- the relationships that bind enterprises into particular ways of doing business (alliances, sub-contracting, franchising, and strategies of conflict avoidance with powerful rivals) (power as relationships); and
- the dominance that some network members can exercise over others, particularly large corporations over sub-contractors at the inter-organisational scale, and headquarters over branch plants at the intra-organisational scale (power as discipline).

In this typology, an initial distinction is drawn between, on the one hand, the large corporations of the corporate sector (including TNCs) and, on the other hand, smaller firms (SMEs). At the *inter-organisational* scale in this segmentation framework, inequalities in power are reckoned to create a dominant, but dynamic and continuously changing core of global corporations surrounded by transnational corporations and other large, multi-location enterprises. Progressing towards the dominated periphery of enterprise networks, a spectrum of legally autonomous smaller firms is identified in the segmentation model of which five types were recognised:

> "... *leaders* (technical and commercial innovators dependent on an individual's drive but prone to financial failure and take-over) ... *loyal opposition* (sometimes small multiplant groups occupying niches in an economy not open to larger business organisations) ... *satellites* (small firms locked into subcontract, franchise, licensing, or agency agreements with larger business organisations ... *satisfied* (small firms kept small by their owners who are reluctant to relinquish personal control) ... *craftsmen* (run by people conversant with making a product but not necessarily with ... commercial flair)." (Taylor 1984, pp.67-68)

However, the larger enterprises in this spectrum are also multi-site and multi-locational operations comprising often large numbers of subsidiaries and associates, branch plants, joint ventures and strategic alliances and so on. *Within* these enterprises, different sub-units have different amounts of power depending upon the internal resources and authority they command. These inequalities in intra-organisational power relations within the corporate sector are viewed in the segmentation model as creating four distinct segments, each of which might be a type of sub-unit or company within a global, transnational or large multi-site corporation. Among these corporate sector sub-units:

> "... [the] most central are the *leading* edge companies that are in the business of creating new products and services. *Intermediates* in this corporate setting are at the heart of the organisation yielding steady, reliable profits from the production of established products. *Laggard* companies mass-produce commodities that are in hyper competitive fields or are fast becoming obsolete. They must continually search for higher productivity to offset shrinking profit margins. Finally, *support* companies provide general services and internalised inputs of commodities to companies of the larger business organisation." (Taylor 1984, p.68)

Recognising and combining these dimensions of inter- and intra-organisational power inequalities, the segmentation framework is, in effect, a caricature of the forms that inequality might take in a capitalist economy - an identification of positions that enterprises might possibly occupy within networks. Implicit in this model is the recognition that inter- and intra-organisational power asymmetries create a lumpy topography of enterprise, with the bunching of business types (the 'statuses' of economic sociology) along continua of inter- and intra-organisational power. How stable that topography might be is difficult to assess since the segmentation model is essentially static, although Taylor (1995) has suggested that Clegg's (1989) concept of circuits of power might be used to breathe a dynamic into it.

To summarise, therefore, the power networks framework proposed here offers a schematic model of the role and functioning of power in shaping trajectories of enterprise development. It combines different views on power with an established model of enterprise segmentation to create an approach to analysis that can be operationalised. There are five essential elements to this approach. First, all business enterprises are seen as operating within a dynamic 'field of forces'. That field involves domination, endowed by work practices and technology, and the codified and tacit relationships that create meaning and membership and locally specific 'ways of doing business' within business communities. Second, the forces themselves constrain the actions of individual enterprises and assign them to 'statuses' (or segments) within network structures. Power, therefore, creates topography of enterprise statuses. Third, the statuses affect the access to resources – information, technology, money and authority - that different kinds of enterprise have. These limitations then impose differential constraints on the potential and the performance of the networked enterprises. Fourth, significantly, those constraints operate not only *between* enterprises (the impact of differential inter-organisational power) but also within enterprises, when those enterprises are multi-plant and multi-locational in make up (differential intra-organisational power). Thus, for example, a TNC branch plant in a developing country might be 'powerful' in its local inter-organisational context, by virtue of the power of the corporation it belongs to, but 'powerless' in its intra-organisational context owing to its subordination within a corporate hierarchy (Clarke 1988). Fifth, enterprises' statuses (and the statuses of plants in large corporations) are enacted and re-enacted on a daily basis, affecting and being affected by their differential and unequal access to resources. That re-enactment reinforces their awareness of inequality generating resentment and resistance and pressure for change. With this model specification it is possible to design a scheme of measurement empirically to test the impact of unequal power relationships of business enterprises' transaction structures. This is the task of the next section of this chapter.

Calibrating 'power' in the power networks model: identifying measurable traits

To operationalise the power networks model, the approach adopted here is to define an enterprise's position or status in the topography of an idealised network in terms of the three forms of power identified in Clegg's (1989) circuits of power model (Taylor 1995). Measures of power are developed to embrace issues of *dominance and subordination* (especially in terms of technology and work regime), *relational inequalities* (enshrined in ways of doing business), and *differential access to resources* (underpinning day-to-day interactions). In addition, the dimensions of inter- and intra-organisational power are operationlised separately. In the single site, single plant SME there is no intra-organisational power differential. The establishment is always the core of the enterprise, even though the enterprise may be powerless in an inter-organisational context. In contrast, a plant in a powerful multi-site, multi-plant corporation could be intra-organisationally powerful or powerless.

Inter-organisational power

To calibrate *inter-organisational power*, five measurable criteria can be identified; transnationality, size, diversity, dynamics and performance (see Clarke 1984,1985,1988). *Transnationality*, or the spatial spread of an enterprise, can be looked on as an index of dominance and empowerment. It involves the control of technologies, the control of opportunities to manipulate transfer pricing flows of all kinds (Taylor 1984), the potential to control sub-contractors and the ability to hedge investment exposure by setting up joint ventures and strategic alliances. *Size*, too, is a significant but complex basis of inter-organisational power (Wrong 1995, McDermott and Taylor 1982). Large size endows an enterprise with effectiveness in shaping inter-organisational relationships and influencing accepted conventions for doing business. It gives an enterprise the power to secure resources and to exercise influence within a network irrespective of its efficiency. The histories of many small firms show only too clearly that efficiency does not automatically equate with effectiveness, especially when it comes to raising and securing capital. Likewise, many large corporations remain effective long after they have ceased to be efficient. *Diversity* in the operations of an enterprise suggests differences in the risk spreading ability of firms and their use of conglomerate strategies and portfolio investment. Such portfolio approaches to business again offer opportunities to firms to hedge against over-reliance on particular production chain relationships.

The *dynamics* of an enterprise are a further source of inter-organisational power, particularly as it affects the securing of resources (especially money and investment) in day-to-day business. Not only do dynamic enterprises have greater power to secure resources, they also have significant relational power, allowing them to set future business agendas in the community as is only too evident in the rise of new Internet businesses in the emerging e-commerce field. Finally, *performance* and especially productivity, endows enterprises with differential inter-organisational power. Day-to-day, it allows them to secure resources, and it supplements enterprise size and dynamics in securing relational power.

Intra-organisational power

To calibrate power in an intra-organisational context involves measuring the unequal control of resources by individual plants and sub-units *within* an enterprise. In Clegg's (1989) terms, the focus here is on power as agency and the exercise of technical and positional power within commercial bureaucracies. The rules of meaning and membership within such multi-site, multi-plant enterprises are established through authority and imposed through bureaucracy. The task for measurement is to find where power lays within the established order of an enterprise. Following Mindlin and Aldrich (1975) and the Aston Group's studies of power, four indices of intra-organisational power can be identified: ownership of equity, positionality, age and autonomy of action. These indices reflect the established relationships of power within an enterprise and their role in securing and acquiring resources.

At its simplest, intra-organisational power finds expression in the *ownership of equity* in a plant, whether it is wholly owned or partially owned. Larger equity holdings in a subsidiary or sub-unit can be equated with greater centrality in the enterprise and, therefore, with greater technical and positional power. *Positionality* can be interpreted as determining how highly or lowly placed a sub-unit is within a bureaucracy – akin to applying the notion of managerial 'spans of control' at the subsidiary level in a corporation. In its crudest form, it is reflected in the proportion of a corporation's employment or subsidiaries controlled from a particular site or sub-unit. More subtlety, it is mirrored in a sub-unit's representation on increasingly higher boards of directors. In both cases, it reflects a sub-units 'substitutability' within the larger enterprise to which it belongs – its positional power in the control of resources. *Age* is a similarly simple index of intra-organisational 'substitutability' and positional power within the established order of a large corporation. There is evidence to suggest that the longer a sub-unit has been part of an enterprise the more likely it is to persist, suggesting that age confers positional power. Certainly, newly acquired establishments are more likely to be closed down when they are taken over and incorporated into larger enterprises, the outcome being the result of well-understood processes of bureaucratic incompatibility.

But, perhaps the strongest index of intra-organisational power is the *autonomy in decision-making* and the discretion over day-to-day operations that a subsidiary or sub-unit within a large enterprise is able to exercise. In classic contingency views, the business organisation comprises a 'core technology' (what the business does or makes) that is surrounded by departments that buffer it from the pressures and uncertainties of the external competitive environment (Lawrence and Lorsch 1967, McDermott and Taylor 1982). Departments are 'boundary spanning structures' and deal with functions such as sale, marketing, purchasing, transport, accounting, personnel and legal matters, for example. Therefore, an index of a sub-unit's intra-organisational power is the number of these functions that it controls itself and the manner in which that control is exercised. Purchases of capital equipment, for example, might be decided at the level of the subsidiary, but those local decisions might have to be ratified elsewhere in the corporations.

The Australian enterprise study

The sample

To assess empirically the impact of power inequalities on firms' transaction structures and interrelationships with other business enterprises, data have been collected from a sample of Australia's engineering enterprises. Businesses were surveyed across the spectrum of enterprise, from parts of TNCs and large national enterprises (subsidiaries and branch plants) to partnerships and small, single plant, family firms. They were surveyed during 1992 in each of four centres, providing a cross-section of types of locality in Australia from a large metropolitan centre to a small and isolated rural town. The centres where business enterprises were surveyed were:

- Perth, one of the Australia's major metropolitan centres – a State capitol - with a significant engineering sector with strong ties to Western Australia's mining industry (sample = 42);
- the Illawarra region on the southern coast of New South Wales which is major coal and steel producing locality centred on Wollongong south of Sydney (sample = 27);
- Albury-Wodonga, twinned settlements on either side of the Murray River in New South Wales and Victoria respectively, which were designated at a growth centre in the early 1970s by the Whitlam Labor government and which has experienced rapid manufacturing growth in the 1980s and 1990s (sample = 9);
- Bunbury, a small and isolated urban centre of less than 30,000 people in the south west of Western Australia (sample = 12).

Measuring inter-organisational and intra-organisational power in the Australian case study

For the present study, the five dimensions of inter-organisational power identified in connection with the power networks model have been measured as nine variables in the present case study. These variables and the dimensions of inter-organisational power they relate to are:

- *transnationality* - measured as (V1) the *numbers of countries* within which an enterprise operates, and (V2) the proportion of those subsidiaries and *sub-units in developed countries*, as opposed to developing countries, reflecting the extent to which their internationalisation goes beyond the global triad economies;
- *size* - measured in three ways, as (V3) *numbers of subsidiaries* and other operating units within an enterprise, as (V4) *total employment* and as (V5) *total revenue*;
- *diversity* - calibrated as the (V6) *numbers of sectors* within which an enterprise operates;
- enterprise *dynamics* - measured as both (V7) *revenue change* and (V8) *employment change*;
- *performance* - constructed as a simple index, (V9) *revenue per employee*.

The four dimensions of intra-organisational power recognised in connection with the power networks model are measured by 20 variables:

- *ownership of equity* - measured as (V1) *the proportion of a sub-units equity held by the parent company*;
- *positionality* – measured as the proportion a sub-unit controls of (V2) *whole enterprise employment* and (V3) *whole enterprise subsidiaries*, and a sub-unit's (V4) *representation on higher boards of directors within a corporation*;
- *age* – (V5) the *number of years* a sub-unit has been part of a particular corporation;
- *autonomy* measured as the extent of local control over 15 boundary spanning functions: (V6) *industrial relations*, (V7) *personnel*, (V8) *maintenance*, (V9) *transport*, (V10) *data processing*, (V11) *purchase of capital equipment*, (V12) *purchase of material inputs*, (V13) *purchase of spares & maintenance*, (V14) *sales & marketing*, (V15) *R&D*, (V16) *advertising*, (V17) *market research*, (V18) *production*, (V19) *accounting*, (V20) *finance*.

Local autonomy in each of the 15 boundary spanning functions is measured on a six-point scale, from '5'(most) to '0' (least);

'5'	on-site, no external ratification needed
'4'	on-site, must be ratified in the group
'3'	bought in from an unrelated supplier
'2'	bought in from an unrelated supplier - must be ratified in the group
'1'	supplied by another member of the group
'0'	function not performed.

The Australian case study: dimensions of inter- and intra-organisational power

Empirically to identify dimensions of inter- and intra-organisational power among Australian engineering enterprises, separate factor analyses have been undertaken using the survey data on inter- and intra-organisational power. The rotated factor loadings from each of these analyses are reported in Tables 9.1 and 9.2.

Table 9.1   Inter-organisational power: rotated factor loadings

Variable	Factor 1	Factor 2	Factor 3
Number of sectors	.795	.434	
% subsides in developed countries	-.780		
Number of subsidiaries	.742	.580	
Number of countries	.703	.617	
revenue per employee	.603		
Employment size 1992		.923	
Revenue 1992		.879	
Employment change 1988-92			.806
Revenue change 1988-92			.800
Percent Variance	32.1%	29.1%	15.3%

Table 9.2   Intra-organisational power: rotated factor loadings

Variable	Factor 1	Factor 2	Factor 3	Factor 4
BS- Industrial Relations	.829			
BS- Personnel	.750			
BS- Maintenance	.662			-.449
BS- Transport	.610			
BS- Data processing	.555			.468
Ownership of equity		.887		
Employment controlled		-.887		
Subsidiaries controlled	.657			
BS- Buying capital equip.		-.604		
BS- Sales and marketing			.780	
BS- R&D			.676	
BS- Advertising			.619	
BS- Market research	.383		.608	
BS- Buying spares/maint.			.421	
BS- Production			.410	
BS- Buying material inputs			.407	
Age				-.653
BS- Accounting			.424	.551
BS- Finance	.410			.510
Board representation				.502
Percent Variance	14.8%	14.4%	14.4%	9.6%

Dimensions of inter-organisational power

The analysis of inter-organisational power reveals three main dimensions that account for more than 75 per cent of the variance in the data set. Factor 1 accounts for almost a third of the variance. It draws together survey firms with the greatest diversity of sectors, the greatest number of subsidiaries, with operations in the greatest number of countries and also the greatest number of less developed countries (i.e. the proportion of subsidiaries in developed counties is negatively loaded). Significantly, these same enterprises are the most productive, with the somewhat crude revenue per employee measure loading positively on the factor (Table 9.1). The sectoral and spatial diversity of these enterprises defines their *global reach* - their transnationality and their integration into the global economy. The largest and most diverse of these enterprises have shed labour and radically restructured in the past decade or more, hence the strong positive loading of the revenue per employee measure on this factor. 'Global reach' gives firms access to resources, continuing influence, and a capacity to dominate.

Factor 2 again explains nearly one third of the variance in the inter-organisational data set. Although it echoes some of the elements of the previous factor in sectoral diversity, numbers of subsidiaries and numbers of countries in which enterprises have operations, it does not include indices of internationalisation. Size measures are the principal traits built into this factor - both employment size and revenue size. Interestingly, this dimension is unrelated to performance as gauged by revenue per employee. This, then, is a factor defining sheer *size* as a source of inter-organisational power that stands in stark contrast to the previously defined dimension of 'global reach'. Size it can be argued sustains effectiveness and maintains dominance amongst business enterprises.

Factor 3 from the inter-organisational analysis accounts for only 15 per cent of the variance and draws together the two key variables measuring *enterprise growth* in the late 1980s and early 1990s - employment growth and revenue growth. Behind global reach and size, the resource-drawing power and influence of growth would appear to be an important basis of inter-organisational power.

Dimensions of intra-organisational power

The analysis of intra-organisational power for Australian engineering establishments identifies four factors relating to aspects of 'power to' and 'power over'. 'Power to' recognising the autonomy of action that can be conferred through an establishment's configuration of boundary spanning structures, and 'power over' reflects the importance of positional variables in the analysis and the hierarchical aspects of control within the bureaucracies of large, multi-location corporations. It is significant that the explanatory power of this analysis is relatively weak. Three of the four factors extracted in the analysis account individually for between 14 and 15 per cent of the variance in the data set, and the fourth factor accounts for less than 10 per cent. The 53.3 per cent of the variance accounted for by the four factors of this analysis contrast sharply with the 75.5 per cent accounted for by the three factors of the inter-organisational analysis.

The variables most strongly loaded on Factor 1 of the intra-organisational analysis are for the local control of industrial relations and personnel functions. Somewhat less strongly loaded on the factor are measures of the local control of production support activities - maintenance and transport. This factor can be interpreted as defining the *local power to control personnel and service inputs*. The weaker loadings for data processing, finance and market research reinforce the service support interpretation of this factor.

Factor 2 combines a very different series of variables. Negatively associated with the factor are the control an establishment has over a whole enterprise's employment and the local control it exercises over capital equipment purchases. Positively associated with the factor, and being inversely related to the apparent directionality of power on this dimension, is external equity control (as might be expected) and also the number of subsidiaries controlled by survey respondents. With this combination of variables, the factor can be interpreted as an index of *positional power* coupled with elements of hierarchical control. On the one hand, the factor distinguishes between single plant enterprises and parts of multi-plant enterprises, at its negative and positive poles respectively. On the other hand, the positive loading of the external equity control variable suggests that, at least in the Australian sample, these controllers of subsidiaries are also controlled. By inference, they occupy the middle tiers of larger corporate structures. The headquarters of those enterprises are elsewhere in Australia or overseas.

Factor 3 is more readily interpreted and combines measures of the local control of sales and marketing, R&D, advertising and market research. Associated more weakly with these output transaction variables are other transactional relationships for the purchase of materials and maintenance inputs, production and associated accounting functions. This combination of variables suggests a measure of autonomy that identifies an establishment's *local power to control external transactions.*

Finally, Factor 4 is the weakest dimension of intra-organisational power explaining less than 10 per cent of the variance in the data set. The strongest contributor to its make up is the negative loading for age. In combination with other positive loadings on this factor, this dimension of intra-organisational power relates to the specific powers of newer corporate subsidiaries that are well connected through representation on higher boards of directors. This gives these enterprises *local power to control accounting and finance* activities (and associated data processing).

Two additional composite measures of inter- and intra-organisational power have been added to the seven dimensions of power generated from the empirical analysis of Australian engineering enterprises. These summary measures have been constructed in the two analyses taken separately by weighting respondents' scores on factors by the variance that factor explains and summing the results for each firm.

Power, segmentation and place

Segmentation

With the empirically derived dimensions of inter- and intra-organisational power for Australian enterprises, it is possible to assess the extent to which Taylor and Thrift's (1982a, b, 1983) segmentation approach to enterprise power remains realistic and relevant in economic geography (Dicken and Thrift 1992, Hayter 1999). There has been only limited research in economic geography on conceptualising the role of power in shaping enterprise relationships and patterns of geographical industrialisation (see Taylor 1995, Allen 1997). Taylor and Thrift's segmentation framework is one of the earliest and most comprehensive. It uses a resource dependence perspective on power (Pfeffer, 1981) to identify and elaborate a typology of enterprise segments. The categories of the typology are 'statuses' reflecting enterprise inequalities in resource control. Corporate sector operations that might be separate subsidiaries within a single TNC, for example, range from 'leaders' (developing new lines of business and profit), 'intermediates' (generating core profits), 'laggards' (extracting residual profits from increasing obsolete activities), and 'supports' (providing administration and specialised internal services). Smaller firms are seen in five more subsets; 'leaders' (innovators), 'loyal opposition' (niche players), satellites (captured by licenses, sub-contracts and franchises), 'the satisfied' (content to retain personal control of their own firm), and 'craftsmen' ('doers' often with little business acumen). In a conceptual form, the subsets of corporate sector operations and smaller firms arranged along a power continuum are shown in Figure 9.1.

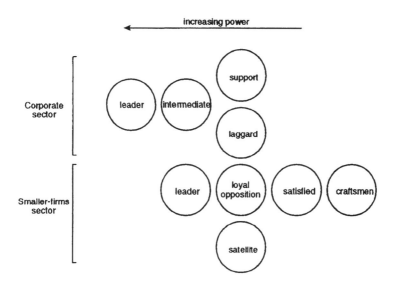

Figure 9.1   Power relationships in the segmented economy

The Taylor and Thrift segmentation model can be criticised for being simplified, static and stylised. In comparison with the more recent views of power outlined in the present paper, the segmentation approach is certainly simplified in that it neglects relational power and power as dominance. These are both forms of power central to the 'circuits of power' approach to inter-enterprise relationships (Clegg and Dunkerley 1980, Clegg 1989, Taylor 1995). Neglecting these aspects of relational power and domination (collectively Clegg's 'field of forces') also leaves the segmentation framework fundamentally static with no stabilising and destabilising pressures to bring about radical and substantive change. The criticism of stylisation comes from simple categorisation of the framework and its idiosyncratic conflation of inter- and intra-organisational dimensions of power. Intra-organisational power and control is identified only in terms of large corporations, while the smaller firms sector is treated as essentially atomistic and single establishment firms. Thus, 'loyal opposition' smaller firms that might themselves operate at multiple locations (Taylor 1983) are denied the possibility of having differentiated internal power structures with significant geographical implications.

The aim of this section is, therefore, to unpack the notion of power in Taylor and Thrift's segmentation framework to show how it can be refined using the fuller 'circuits of power' concepts derived from Clegg (1989) that are used in the analyses of this paper. To show how segmentation relates to the dimensions of power developed here it is first necessary to allocate the surveyed Australian engineering firms to the segments of the Taylor and Thrift model. This allocation has been undertaken using the decision tree in Figure 9.2, derived from Taylor and Thrift (1982b) and Taylor (1984). Here key characteristics that are recognised as dividing one segment from another are arranged so survey respondents can be cascaded through the decision points to assign them unambiguously to a segment. To assign respondents to the corporate sector segments, the decision rules (in boxes in Figure 9.2) identify multi-plant, non-independent operations with five or more establishments. These, in turn are divided on whether; (1) they are group, regional or product headquarters locations, (2) on-site employment is over 100, (3) R&D employees make up more than 11 per cent of the labour force, or (4) the vast majority of the labour force are operatives. The assignment rules for the smaller firms segments are again based on (1) employee numbers, (2) operative numbers, and (3) the on-site presence of R&D, but with a significant addition of the extent to which output is controlled by the enterprise's largest customers. The threshold values used in these assignment rules are to a degree arbitrary, but they reflect empirical regularities recognised in studies of the UK engineering industry (Taylor 1983), the UK ironfoundry industry (Taylor and Thrift 1982b) and the Scottish electronics industry (McDermott and Taylor 1982).

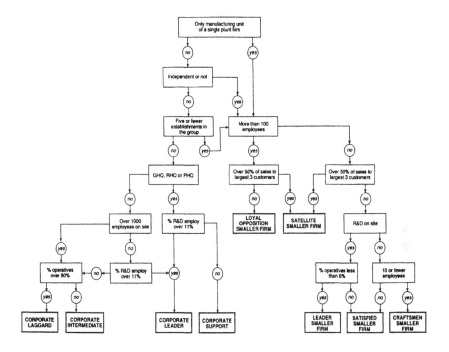

Figure 9.2   Enterprise segmentation: allocation decision tree

One-way analysis of variance has then been used to assess the strength and direction of relationships between enterprise segmentation and the dimensions of inter- and intra-organisational power developed in this study. The results of these analyses are reported in Table 9.3.

The analyses demonstrate a strong positive relationship between enterprise centrality within the segmentation framework and inter-organisational power measured *global reach* and *size*, with F statistics significant at better than the 0.1 per cent level. As might be expected from a static model such as the segmentation framework, the network positions it identifies do not mirror the *growth* dimension of inter-organisational power differentials. Nevertheless, the compound of all three dimensions of inter-organisational power measured for the Australian engineering enterprises – the inter-organisational composite index - is very strongly associated with the pattern of segmentation. The form of this relationship is shown in Figure 9.3.

Table 9.3   Power, segmentation and place: significant relationships
(F statistics with probabilities in parentheses)

Dimension of Power	*Enterprise Segment*	*Place*
Global reach	17.98(0.00)	1.11(35.0)
Size	5.57(0.00)	1.32(27.3)
Growth	1.59(15.1)	0.40(75.0)
Inter-organisational composite index	38.82(0.00)	0.79(50.3)
Local control personnel & service inputs	2.41(2.70)	0.35(79.2)
Positional power	30.19(0.00)	0.50(68.5)
Local control of external transactions	1.71(11.7)	1.37(25.8)
Local control of finance and accounting	4.01(0.10)	0.10(96.1)
Intra-organisational composite index	8.89(0.00)	0.64(59.4)

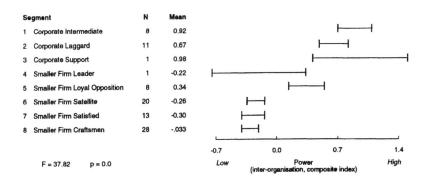

Figure 9.3   Inter-organisational power (composite index) and enterprise segmentation

At the intra-organisational scale, the relationship between the categories of segmentation and the measures of intra-organisational power developed here is again reasonably strong but also somewhat more complex (Table 9.3). The general tendencies in the data are most clearly illustrated in the context of positional power (Figure 9.4). Here it is clear that the most powerful are the corporate sector intermediates, laggards and supports. All the smaller firms segments, with the exception of the loyal opposition, clearly have less intra-organisational positional power, as would be expected. The surveyed loyal opposition is, in fact, more akin to their corporate sector counterparts on this dimension of power. This finding highlights the previously recognised limitation of the segmentation model brought about by its conflation of the quite distinctive aspects of inter- and intra-organisational power. However, there are still very obvious and strong resonances between the present analysis and the segmentation model that it unpacks. The segmentation framework mirrors power inequalities in terms of local control of personnel and services and the local control of finance and accounting. However, it is particularly significant for the present analysis of enterprise transaction structures that the segmentation model only very weakly connects with issues surrounding the local control of external transactions in business enterprises (F=1.71). Nevertheless, from the strength of the relationship between the segmentation framework the composite index of intra-organisational power developed in this study it has to be concluded that, overall, the segmentation model is a reasonable general approximation of patterns of intra-organisational power differentials.

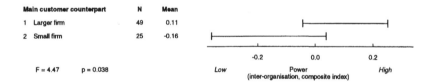

Figure 9.4   **Inter-organisational power (composite index) and largest customer counterpart**

Place

A further important geographical research question arises from the close relationship between enterprise power and segmentation. The question concerns the extent to which inter- and intra-organisational power relationships are related to and associated with the 'place' characteristics of where firms are located. The Australian data allow this issue to be addressed in at least an exploratory way.

As outlined earlier, Australian engineering enterprises were surveyed in four very different types of place: (1) a major metropolitan centre (Perth, WA); (2) a major coal and steel producing locality (Wollongong, NSW); (3) an inland, formerly designated growth centre (Albury-Wodonga, NSW/Victoria); and (4) a small isolated urban centre (Bunbury, WA). Each of these places has a very different type of economy that is more or less resilient to change. These differences have been demonstrated by the Australian Federal Government's Office of Local in an index of vulnerability it has produced to reflect regions' susceptibility to the "adverse impacts of sustained structural change" (OLG 1992, p.8). Ranked over 94 regional units, from '1' (high) to '94' (low), the four regions' ratings were; Wollongong 22, Albury-Wodonga 30, Perth 30, and Bunbury 81.

One-way analysis of variance has been used to explore whether there is greater within region rather than between region differences in enterprise power differentials. The results of the analysis are very clear cut and are reported in Table 9.3. In contrast to the relationship between power and segmentation, there are no significant relationships at all between the dimensions of inter- and intra-organisational power elaborated here and 'place' calibrated as the four centres from which respondents were drawn. The suggestion is, therefore, that differences in enterprise power are structure-based rather than place-based. This finding holds clear implications for the processes of place-based learning identified as the drivers of industrial districts, regional innovation systems and new industrial spaces. Place may not be as important as these models suggest, but this is an issue that warrants much fuller analysis.

The impact of power on transaction structures

Against this background, a question arises about the impact that unequal power relationships have on firms' transaction structures.

To match the nine measures of inter- and intra-organisational power developed for the surveyed firms, nine variables have been constructed to describe the nature and form of their input and output transactions. The nine variables describe five important aspects of firms' transactions. First, they consider the *numbers of buyers and sellers* a firm has to provide some indication of the size of the transaction network within which they operate. Second, the *concentration of transactions* is taken into account to show the extent to which the survey respondents are captured capitalists dependent on very few suppliers and customers and, potentially, pressured by monopolists. Third, the *local orientation of transactions* is measured as an indicator of firms' local and locational orientation (that can alternatively be interpreted as a measure of parochialism). Fourth, the *contract type* under which they deal with buyers and suppliers is assessed. The issue here is to measure how arms-length firms' transactions are. The weakest contracts are single orders. Contracts become stronger through single contracts with tenders, short-term contracts of less than one year with draw-off arrangements and then long-term contracts of more than a year with draw-off arrangements. Common ownership, when a firm's suppliers and/or customers are members of the same group as itself, can be considered the strongest contractual tie (i.e. the strength of intra-firm trade). Fifth, the *nature of the counterparts* in transactions is considered to provide an indication of the extent to which respondents are linked into corporate sector as opposed to smaller firms sector networks. Being linked into the former has been hypothesised as involving subordination, while being linked into the latter has been seen as a prerequisite for local learning and the exchange of tacit information.

The nine variables constructed separately for customer and supplier transactions are:

- the numbers of customers (V1) or suppliers (V10) firms have;
- the proportion of their output (by value) that is sold to (a) their largest (V2), and (b) their largest-three customers (V3);
- the proportion of their inputs (by value) that comes from (a) their largest (V11) and, (b) their largest-three suppliers (V12);
- whether their main customer (V4) or supplier (V13) are local, i.e. in the same centre as themselves;
- the number of their five main customers (V5) or five main suppliers (V14) that are local;
- the contract type under which they trade with their main customer (V6) or supplier (V15) (the scale ran from '1' weakest to '5' strongest – '1' single orders, '2' single contract with tender, '3' short-term contract (less than one year) with draw-off arrangements, '4' long-term contract (more than one year) with draw-off arrangements, '5' common ownership with the customer or supplier);
- the predominant contract type among their five main customers (V7) and five main suppliers (V16);
- whether their main customer (V8) or supplier (V17) is a small firm (self defined);

- whether their five main customers (V9) or five main suppliers (V18) are predominantly small firms.

One-way analysis of variance has been used to estimate the strength of the association between the extent of the unequal inter- and intra-organisational power that Australian engineering firms wield and the nature and form of their transaction structures. The results of these analyses are reported separately for sales transactions in Tables 9.4 and 9.5 and for purchasing transactions in Tables 9.6 and 9.7.

Power and the nature of sales transactions

For the impact of *inter-organisational* power on firms' sales transactions, a strong and consistent set of relationships emerges from the analyses of variance reported in Table 9.4. First, it is clear that all dimensions of inter-organisational power have a pervasive effect on nearly all aspects of engineering firms' sales transactions. The only exception is for the total numbers of customers they have. 'Global reach' and 'size' affect sales concentration and sales orientation: the proportion of output sold to their largest customers and whether those customers are 'local' or not. Also, while 'global reach' impacts on whether sales counterparts are small firms or not, whole enterprise 'size' affects the nature of sales contracts. The 'growth' dimension of inter-organisational power has a similar, though less pervasive effect on customer orientation and customer contract types. Second, these relationships compound in the composite index of inter-organisational power to show clearly that inter-organisational power shapes the concentration, spatial orientation, contractual arrangements and counterpart types in engineering firms' sales transactions.

Third, the details of the analyses show that these impacts occur in a consistent direction;

- the more powerful the enterprise to which a respondent belongs (sub-unit of a corporation or and SME) the less locally orientated are its sales;
- the more powerful the enterprise to which a respondent belongs the less likely are its customer counterparts to be small firms;
- the weaker the enterprise the more likely are its contracts with customers to be either (a) single orders and single contracts, or (b) long-term contracts with draw off arrangements; and
- the more powerful the enterprise the more likely are its contracts with customers to be short-term contracts or for sales to be locked into channels of intra-firm trade.

Table 9.4   Sales transactions and inter-organisational power: one-way analyses of variance
(F statistics in bold, probabilities in parentheses)

Transaction characteristics	Inter-organisational Power			
	F1 Global Reach	F2 Size	F3 Growth	Composite Index
Number of customers	*	*	*	*
% sold to largest customer	*	*	*	*
% sold largest 3 customers	2.20 (.10)	2.69 (.06)	*	2.78 (.05)
Main customer 'local'	6.06 (.02)	7.58 (.01)	3.20 (.08)	20.03 (.00)
No. of 5 main customers 'local'	4.06 (.00)	2.26 (.06)	*	8.95 (.00)
Main customer contract type	*	5.77 (.00)	*	4.43 (.00)
Predominant custom. contract type	* (.00)	5.82 (.05)	2.71	5.06 (.00)
Main sales counterpart 'small firm'	*	*	*	*
Sales counterparts predominantly 'small firms'	6.83 (.01)	*	*	9.84 (.00)

The details of these impacts of inter-organisational power on sales transactions can be illustrated with two examples: the impact of power on counterparts and contract types. Figure 9.3 shows graphically the tendency of the inter-organisationally weak to sell to small firms. Figure 9.4 shows the more complex relationship between inter-organisational power and the nature of the contracts under which sales are made to respondents' principal customers. Here there is a strong suggestion that the weakest enterprises are either marginalised, through short-term selling arrangements, or are locked-in as captured capitalists, through long-term contacts with draw-off arrangements. Equally, some sub-units of more powerful enterprises sell through shorter-term contracts with draw-off arrangements. In the terms used here, these operations are neither 'marginalised' nor 'captured'. Other parts of powerful enterprises are, however, bound up in intra-firm trade. These contractual characteristics have all the hallmarks of the operations of corporate branch plants rather than corporate control centres, a major issue in a country like Australia with extensive foreign direct investment.

Table 9.5 Sales transactions and intra-organisational power: one-way analyses of variance
(F statistics in bold, probabilities in parentheses)

Transaction characteristics	Intra-organisational Power				
	F1 Person-nel	F2 Centr-ality	F3 Trans-actions	F4 Fin. & Acctng	Composite Index
Number of customers	*	*	*	*	*
% sold to largest customer	*	*	*	*	*
% sold to largest 3 customers	*	*	*	*	*
Main customer 'local'	**4.53** (.04)	**6.58** (.01)	*	*	*
No. 5 main customers 'local'	*	**3.22** (.01)	*	*	*
Main customer contract type	**2.36** (.06)	**2.56** (.05)	*	**2.17** (.08)	*
Predominant custom. contract type	**3.78** (.02)	**3.34** (.03)	*	**5.39** (.02)	*
Main sales counterpart 'small firm'	**7.89** (.00)	*	*	*	*
Sales counterparts predominantly 'small firms'	*	**6.42** (.01)	*	*	*

As it is measured here, the impact of *intra-organisational power* on sales transactions is much less pervasive than the impact of inter-organisational power (Table 9.5). First, 'local control of personnel' is the principal dimension of intra-organisational autonomy in the present study (Factor 1) and, together with 'centrality' (Factor 2) it is consistently associated with the nature of respondents' ties with customers. These two dimensions of power affect the spatial orientation of sales, the form of sales contract and the type of customer counterpart (Table 9.5). As might be expected, 'local control of finance and accounting' (Factor 4) also impacts at this intra-firm scale on customer contract types.

Second, and at first sight quite contrary to the relationships for inter-organisational power, the more intra-organisationally *powerful* a respondent:

- the more *local* are its sales transactions;
- the more likely are contracts with customers to be single orders/ single contracts or long-term contacts – echoing the duality of *'marginalisation'* and *'capture'* through contractual arrangements on sales identified for the *least* powerful in the inter-organisational context;
- the more likely are its customers to be *small firms*.

These apparently perverse results in fact reflect the nature of the measurement system used in this study. Small, single plant firms by definition have boundary spanning functions 'on-site'. The bulk of the intra-organisationally 'powerful' in the survey are, therefore, the single plant firms, and the intra-organisationally 'powerless' are corporate sector branch plants. Under these circumstances, these weaker 'branch plants' appear to have transaction structures that reflect the nature of the larger corporations to which they belong, i.e. reflecting their inter-organisational *powerfulness*, not their intra-organisational *powerlessness*. This finding also suggests significantly, but only very tentatively, that inter-organisational power relationships are more important than intra-organisational power relationships in shaping firms transaction structures. If this finding holds true, it calls into question the likelihood of any TNC or corporate sector plant being locally embedded in host country communities.

Third, in comparison to the analysis of inter-organisational power, the dimensions of intra-organisational power do not combine to create a strong composite index. Instead, they appear to confound each other producing no pattern whatsoever.

Power and the nature of transactions with suppliers

The relationships between power and the purchasing transactions of Australian engineering enterprises mirror, though far less strongly, those already revealed for sales transactions. The significant statistical relationships revealed by one-way analysis of variance are reported in Tables 9.6 and 9.7. For *inter-organisational power*, respondents belonging to enterprises with the greatest 'global reach' tend to have the most suppliers and this relationship is reflected in the composite inter-organisational power variable. Greater 'size' as a basis of power is, once more, directly related to respondents having progressively more non-local suppliers. Also, increasing 'global reach' and 'size' are directly related to inputs being bought through progressively stronger forms of contract, except again for the weakest being tied into long-term supply contracts. It is, in addition, the least powerful in terms of 'global reach' that again tend to buy from small firm counterparts. These relationships are demonstrated graphically in Figures 9.6 and 9.7, for (1) the inter-organisational 'composite index' and 'numbers of suppliers' and (2) 'global reach' and 'predominant contact type' respectively. While Figure 9.6 displays a simple, direct linear progression, Figure 9.7 displays the more complex 'marginalisation-cum-capture' relationship already discussed for contractual arrangements on sales.

Finally, for *intra-organisational power* and purchasing, three simple relationships are demonstrated in Table 9.7, and these again mirror the relationships for intra-organisational power and sales. First, declining 'centrality' (Factor 2) *within* an enterprise, i.e. among sub-units with the least positional power in a business organisation, tends to be associated with them having greater numbers of suppliers. Second, declining 'centrality' is also associated with respondents having stronger supplier contracts. Third, those with the least on-site control of finance and accounting (Factor 4) tend to buy from corporate rather than small firm counterparts. All these relationships are consistent with the nature of business operations in peripheral corporate branch plants, distinguishing them from other parts of large corporations and single plant small firms where, by definition, all activities are centralised on one site. Here again is confirmation of the apparent primacy of inter-organisational power relations over intra-organisational power relation, but in this case in terms of purchasing transactions. Once more it throws doubt on the potential for corporate sector plants to be locally embedded.

Table 9.6 Purchasing transactions and inter-organisational power: one-way analyses of variance
(F statistics in bold, probabilities in parentheses)

Transaction characteristics — Inter-organisational Power

Transaction characteristics	F1 Global Reach	F2 Size	F3 Growth	Composite Index
Number of suppliers	**7.08** (.00)	*	*	**2.97** (.02)
% bought from largest customer	*	*	*	*
% bought from largest 3 suppliers	*	*	*	*
Main supplier 'local'	*	**3.58** (.06)	*	*
No. of main 5 suppliers 'local'	*	*	*	*
Contract type with main supplier	**6.51** (.00)	**3.08** (.00)	*	**5.91** (.00)
Predominant supplier contract type	**6.80** (.00)	**4.45** (.03)	*	**7.49** (.00)
Main supplier a 'small firm'	*	**2.29** (.09)	*	*
Suppliers mainly 'small firms'	*	*	*	*

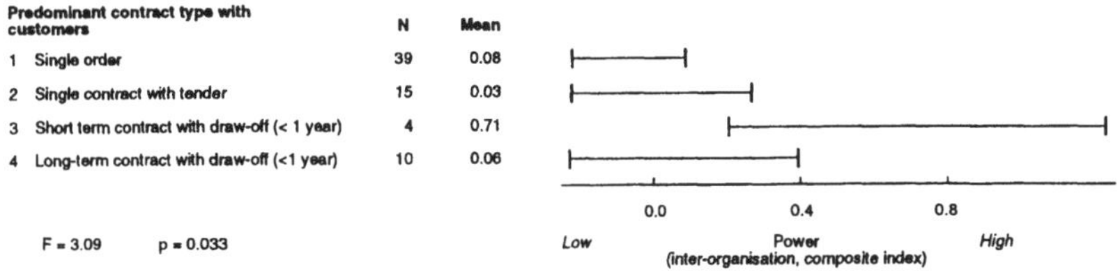

Predominant contract type with customers	N	Mean
1 Single order	39	0.08
2 Single contract with tender	15	0.03
3 Short term contract with draw-off (< 1 year)	4	0.71
4 Long-term contract with draw-off (<1 year)	10	0.06

F = 3.09   p = 0.033

Low   Power (inter-organisation, composite index)   High

Figure 9.5 Inter-organisational power (composite index) and predominant contract type with customers

Table 9.7   Purchasing transactions and intra-organisational power: one-way analyses of variance
(F statistics in bold, probabilities in parentheses)

Transaction characteristics	Intra-organisational Power				
	F1 Person-nel	F2 Centr-ality	F3 Trans-actions	F4 Fin. & Acctng	Composite Index
Number of suppliers	*	**3.32** *(.01)*	*	*	*
% bought from largest supplier	*	*	*	*	*
% bought form largest 3 suppliers	*	*	*	*	*
Main supplier 'local'	*	*	*	*	*
No. of 5 main suppliers 'local'	*	*	*	*	*
Contract type with main supplier	*	**8.84** *(.00)*	*	*	**3.53** *(.01)*
Predominant supplier contract type	*	**6.78** *(.00)*	*	*	**2.17** *(.08)*
Main supplier a 'small firm'	*	*	*	**5.77** *(.02)*	*
Suppliers mainly 'small firms'	*	*	*	*	*

Number of suppliers		N	Mean
1	1 - 25	17	-0.19
2	26 - 50	23	-0.18
3	51 - 100	26	-0.02
4	101 - 200	11	0.06
5	>201	6	0.74

```
                        0.0         0.5         1.0
            Low                   Power              High
                       (inter-organisation, composite index)
```

Figure 9.6   Inter-organisational power (composite index) and number of suppliers

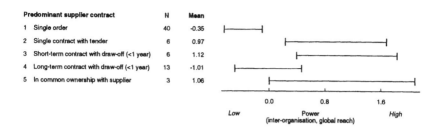

Predominant supplier contract	N	Mean
1 Single order	40	-0.35
2 Single contract with tender	6	0.97
3 Short-term contract with draw-off (<1 year)	6	1.12
4 Long-term contract with draw-off (<1 year)	13	-1.01
5 In common ownership with supplier	3	1.06

Figure 9.7   **Inter-organisational power (global reach) and predominant contract type with suppliers**

Discussion and conclusions

This chapter began by contending that current research on new industrial spaces, learning regions and regional innovation systems has ignored the role of power inequalities in shaping relationships between firms, especially for buying and selling. To explore this issue, power has been conceptualised as 'circuits of power', and a scheme of empirical measurement has been devised. Empirical analyses based on this framework and methodology applied to survey data for Australian engineering firms significantly unpack the nature of enterprise power inequalities and substantially refine Taylor and Thrift's (1983) earlier model of unequal power relationships based on a crude typology of enterprise segments.

    What is clear from the analyses in this chapter is that power plays a pivotal role in shaping the form and nature of firms' buyer/supplier transactions. Power asymmetries are systematically associated with:

- the numbers of suppliers firms have;
- the spatial orientation of their sales and purchases;
- the nature of the contract types used for buying and selling; and
- the extent to which firms' transactions are with small firms as opposed to corporate counterparts.

The analyses also suggest that enterprise power asymmetries are unrelated to place, with the implication that power differentials are institutionally based not place-based.

For inter-organisational power relationships the same consistent patterns hold true for both selling and buying:

- *powerlessness* is associated with both strong local linkage and trading with small firm counterparts;
- *powerfulness* is associated with firms having spatially wider transaction and those transactions being more frequently with corporate sector counterparts;
- *powerlessness* involves either weak contractual relationships or very strong contractual relationships for both buying and selling – either the 'marginalisation' or the 'capture' of firms in buyer/supplier networks;
- *powerfulness*, in contrast, involves either shorter-term contracts (providing certainty but not lock-in) or the absorption of an enterprise's transactions into intra-firm trade.

For intra-organisational power, a complementary pattern emerges. The *intra*-organisationally weak - the powerless in this context - have the same transaction forms as the *inter*-organisationally strong - the powerful in a very different context. They have the least locally orientated buyer and supplier transactions, the fewest transactions with smaller firms, and contractual arrangements that in some cases tie them in to intra-firm trade, but does not marginalise them by doing business only on single orders. The intra-organisationally weak are corporate sector subsidiaries and branch plants. The pattern relationships between power and transactions revealed here suggests that corporate subsidiaries and branch plants have patterns of transactions determined not by their own characteristics, reflecting their intra-organisational subordination, but by the characteristics of the larger corporations to which they belong – their 'global reach', 'size', and 'growth'.

These findings hold important implications for models of local growth based on local embeddedness, local learning processes and local innovation systems. Because of their powerfulness or powerlessness, the empirical evidence of this chapter suggests that business enterprises are incorporated into local economic systems in very different ways, affecting the nature, form and orientation of their transactions. By extension, it can be suggested that these power asymmetries will similarly affect the nature and form of their information flows, and thus the volumes and flows of information within networks. As such, power will affect the incorporation of firms into and the exclusion of firms from local processes of learning.

Not every enterprise in a place will be able to participate to the same extent in the learning process.

But, the results of the present analysis also suggest that the least powerful enterprises, the smaller firms, tend to buy and sell from the equally powerless. Those transactions may be localised, consistent with learning processes, but significantly they are forms of transaction that either marginalise small firm through one-off transactions or 'capture' them and lock it into long-term transactions. Neither of these forms of transaction would appear consistent with reciprocal exchange relationships in socially embedded network systems. Indeed, if information flows are as asymmetric as power relations, then it can be conjectured that in terms of learning the powerless will be givers and the powerful will be takers. Thus, the information arbitrage of large corporations can serve to siphon information from a place at the expense of that place.

The analyses of this chapter also suggest that plants and subsidiaries of corporate sector enterprises have few local transactions, and tend to deal with corporate rather than small firm counterparts for both buying and selling. In other words, irrespective of their information siphoning potential, their embeddedness in small enterprise systems, in Australia at least, is relatively minimal. By inference, they are not central participants in local transaction networks and local learning systems. This is a finding very much at odds with the TNC embeddedness thesis advanced by Dicken (1994) and Yeung (1998a, 1998b). It suggests that the over-enthusiastic extension of local collaborative learning processes to corporate sector enterprises needs urgently to be tempered by recognising the negative effects of power inequalities.

Clearly the present analysis has been exploratory. What it has demonstrated, however, is the need to build more fully into analyses of the functioning of local enterprise systems the impact of unequal power relationships between firms. Power asymmetries have the potential to divide and exclude. They have the potential to impede and damage local learning processes. Only when we more fully understand these ramifications of power in enterprise systems will we begin to appreciate the triggers that can cause some places to decline and others to rejuvenate and grow. Without this understanding, models of local growth built on network processes of learning and embeddedness offer a very imperfect foundation on which to build policies to foster local economic development.

10 Explaining Internationalisation of SMEs: The Importance of Internal and Local Resources

Heikki Eskelinen and Eirik Vatne

Introduction [65]

The implications of the international exchange of goods and services for territorial differentiation and the division of labour have long been a prominent subject in economic geography. In general, the importance of international trade is of more profound interest in small countries. For example, the Nordic countries (Denmark, Finland, Iceland, Norway and Sweden) are small open economies with greater trade exposure than the large OECD countries.

In line with the principle of comparative advantage, the export industries of the Nordic countries have developed around the exploitation of natural resources such as land, minerals, forest, fish, petroleum and hydropower. The drive for a dynamic competitive advantage by the leading firms has supported the evolution of related industries. The emphasis of mass or process technology, large-scale operations, global markets and large company size has been typical features of these Nordic clusters (see, Sölvell et al 1991).

However, the rapid growth of intra-industry trade is an important characteristic of the ongoing industrial transformation. It is partly due to the sophistication and complexity of manufacturing industry in developed countries, which derives from the search for flexibility and the evolution of a network economy in a post-Fordist period. The development of intra-industry trade has opened up export markets for smaller firms operating in specialised niches.

This chapter deals with the internationalisation of Finnish, Swedish and Norwegian small and medium sized enterprises (SMEs). The analysis derives its impetus from the question of whether Nordic SMEs will be able to stay competitive in markets which will become increasingly internationalised as a result of the EU internal market and other ongoing changes.[66]

In the debate concerning the future prospects for Nordic SMEs, and for Nordic industry in general, the geographical location and spacio-economic characteristics of these countries have received a great deal of attention. Distinctive Nordic features include low population densities, scattered settlement structures and high transport costs. In comparison to their competitors in Western Europe, Nordic SMEs are located far from the core of the internal market, usually in small communities.

The characteristics of a local operational environment are of particular importance to a small firm in its demand for resources, which are scarce by definition. An SME has only limited possibilities of influencing its business environment, and it has to rely on external resources in its strategic activities, for instance in internationalisation. If, as can be assumed, the quality, accessibility and costs of the required resources are dependent on the structural features of the community or region concerned, SMEs closer to the main markets in bigger and more diversified centres tend to strengthen their competitiveness as the internationalisation of economic activities proceeds.

Barriers to SME export

Resource problems

Earlier research on internationalisation and export behaviour has shown different attitudes between SMEs and larger enterprises (Cavusgil and Nevin 1981). Larger companies put more emphasis on marketing in general and focus their marketing more specifically towards the needs of international customers (Sriram and Sapienza 1991). Inside the SME group resource rich (larger) enterprises seem to form a more active export policy than poorer (smaller) enterprises (Bijmolt and Zwart 1994). On the other hand, many SME exporters seem to act reactively in their exporting operations (Kaynak, Ghauri, Olofsson-Bredenlöw 1987). Very often they lack general knowledge of foreign markets, have problems in judging their competitive stand in international markets and seem to have trouble in developing a focused export strategy (Edmunds and Sarkis 1986, Green and Larsen 1987, O'Rourke 1985). Thus, a shortage of export-related information seems to hinder many SMEs from exporting (Reid 1984).

SMEs also reportedly face special challenges caused by a more stressed financial situation with low equity and poorer profits. They tend to have more problems than larger firms in adjusting to foreign or international laws and industrial standards, a shortage of business connections and international networks, and difficulties in organising adequate distribution (Bilkey and Tesar 1977). Many of these problems are connected to a lack of critical mass in SMEs (Attiyeh and Wenner 1979). Turnover, value added and profits per capita are also generally lower in SMEs (Rothwell and Zegveld 1982). SMEs also reportedly have a shortage of international marketing specialists in their organisations (Dichtl, Leibold and Mueller 1990, Diamantopoulos 1988).

Earlier research also provides evidence that SMEs suffer characteristic management problems. The failure rate among SMEs is high. As a matter of fact, many small firms do not want to grow (Penrose 1959), but sheer survival is often their day-to-day strategy (Boswell 1973). Neither do small firm entrepreneurs think strategically and, consequently, fail to anticipate and avoid threats. Even if they detect threats in time, they still lack personnel with the intellectual capacity and material resources needed to handle such challenges efficiently (Hull and Hjern 1987). In many smaller firms imitation is therefore more important to business behaviour than formal planning.

In the light of the above findings, it is understandable that SMEs have not been able to expand their activities into international operations in the same manner as larger corporations. Christensen (1991) claims that two central factors differentiate SMEs from larger firms in this respect. First, SMEs as organisations are generally less committed to exporting and other international operations. Second, size seems to differentiate the ways firms collect and process market information. Larger firms more easily formulate a purposeful strategy for information processing and marketing whereas smaller firms seem to be more influenced by contingent events and experiments.

Management

Obviously, the general scarcity of human and material resources in SMEs is an important explanation of why smaller firms are less committed to exporting than larger ones. A major aspect of this is the quality of the manager. The entrepreneur often dominates smaller firms (Imai and Baba 1991). Their advantage, but also disadvantage, is their multiple control over the firm, ranging from shareholding to direct inspection on the shop floor. The manager's personal characteristics, such as education, work experience, social network and personality are therefore strongly intertwined with the organisation of the firm (Bouwen and Steyaert 1990).

In sum, the manager's personal resources are clearly crucially important for SMEs. The ability of an entrepreneur to establish and develop social relations with individuals in other firms is of utmost importance in the use of inter-firm relations. Well functioning and trustful social relations give access to external resources, to information not available in an open market, and to specific fields of know-how for upgrading the capability of the SME concerned. At the same time, the dominant position of the manager/entrepreneur can also be one of the main reasons for the mismanagement of an SME (Rainnie 1989).

Access to external resources

As economic organisations SMEs are continuously confronted by threats to their survival in normal competitive markets. In general, the survival of an organisation is dependent on its ability to acquire and maintain the resources needed to respond to environmental changes. The ability to reduce environmental uncertainty reflects the effectiveness of a firm (Pfeffer and Salancik 1978).

Other aspects of the firm-environment interface have been analysed in theories on transaction costs, social exchange and strategic management. These research traditions have had a decisive impact on the empirical setting of the present study (Christensen et al 1990, Christensen and Lindmark 1993, Vatne 1995). In the following, only some arguments are briefly summarised.

A firm internalises activities as long as the cost of carrying out a transaction internally is less than using an exchange on the open market (Coase 1937). According to this view, whether a transaction is internalised or takes place as an external operation depends on the nature of the transaction involved. Given transaction-specific assets of a semi-specific nature, the transaction cost theory predicts that a transaction should be governed under bilateral/trilateral agreements in inter-organisational arrangements where contracts safeguard the interests of the parties involved (Williamson 1979, 1985). In this sense, new institutional economists claim that trust among partners can be institutionalised in a market economy through explicit or implicit contracts.

There is also a strong interest in the organisation-environment interface in the strategic management literature (Porter 1985). According to this literature, assets of high specificity such as natural or technological resources, human assets and know-how represent the strategic core of a firm. Internal resources and management are concentrated on sustaining and developing these core skills. In order to extract value from them, the firm has to link up with suppliers of material, components and services.

These supplied skills can be embodied or disembodied in character. Their intangibility is often argued to be the reason for firms being actively involved in inter-organisational relations, preferring vertical co-operative arrangements to vertical integration. In this context, the question of ownership can be of secondary importance because vertical control and lower uncertainty is what matters from a strategic point of view (Reve 1990).

In opposition to Williamson's (1985) claim that contracts can be a substitute for trust, Granovetter (1985) defends a behavioural model and insists on an active role for personal contacts and structures in developing dynamic inter-firm relations. Such a model is based on the assumption that a particular value system is linked to these relations, the identification of partners being a critical factor in its formation. Social relations are established between individuals and tend to be long lasting. On the other hand, firm relations based on contracting will often tend to be more short-sighted in character.

Overall, developing interfirm relationships is often a cumulative process of adjustment, investment and development of mutual trust, bonds and dependence. The adaptation process often advances as interactive learning and enables firms jointly to create intangible cross skills, which are difficult to imitate (Johansson and Mattsson 1987). According to social exchange theory, social relations evolve slowly, often starting with minor episodes and transactions, which require only little trust and risk. Over time both parties can demonstrate increasing commitment and trustworthiness (Blau 1968).

The so-called network approach to international marketing has dealt with inter-organisational issues, placing special emphasis on networking as a way of acquiring market intelligence, developing distribution channels and creating innovative systemic products for foreign markets (Johansson and Mattson 1987). Yet these studies have only focused on the trajectory of the firm/market interface. As Christensen (1991) points out with reference to Porter (1985), the internationalisation process of a firm should not only be seen in the light of its horizontal position in a value-added chain, but must also be related to its vertical position and to the interface between its internal and external support systems.

Social and geographical proximity

Given the relevance of social and cultural conditions in the creation of synergetic interrelationships, proximity can be assumed to be of importance in developing inter-firm links. Its role is especially accentuated in the case of the interrelationship concerned being preoccupied with extensive dialogue to solve open-ended problems (Lundvall 1988). This is a common setting for SMEs, which are often involved in the production of customised products in intra-industry transactional arrangements where extensive communication is needed.

In the case of SMEs in particular it can be argued that not only social but also spatial proximity to supplied skills is of importance in enabling these firms to take part in interfirm arrangements and to develop these relations into dynamic and innovative exchanges (Harrison 1992). Smaller firms are embedded in local production conditions through factors ranging from the skills and education of the manpower available, R&D institutions, communication facilities, and public support systems to the industrial environmental texture.

As discussed other places in this book, theoreticians of 'industrial districts' explain the dynamics of spatially concentrated SME-dominated industrial growth by the social and geographical proximity resulting in an extensive division of labour. A contingent historically based entrepreneurial culture supported by the availability of external economies seems to be among the distinctive structural features of industrial districts. Yet, there are also many transactions which do not require proximity. Standardised operations with less need for dialogue and exchange of information can be safeguarded by routines and contractual agreements and can therefore be fairly easily performed over long distances (Scott 1988a). For SMEs, with their shortage of internal resources and their embeddedness in a local business milieu, the role of social relations is probably emphasised in relation to formalised contractual agreements. Assuming the special importance of trust for an SME, the proximity to business partners tends to grow in importance both in horizontal and vertical relations.

Model for empirical investigation

From the above discussion it can be concluded that SMEs with limited internal resources are strongly dependent on their environment and on access to external resources in order to supplement their own limited resource base. The successful development of an SME presumably demands the specialisation and concentration of its core skills and the use of external resources in the execution of specific production tasks, in services or in distribution and marketing. The entrepreneur/manager plays a crucial role in establishing and developing contacts with external partners, especially in dynamic inter-firm relations based on the trustful exchange of sensitive information and involving asset specific investments.

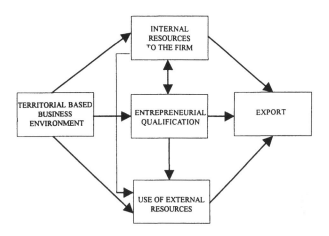

Figure 10.1   A conceptual model of the internationalisation of SMEs

The dependence of dynamic relations on social exchange and the need for trust also put proximity and location into focus. A resourceful business environment gives better and cheaper access to important information on technological or market development, and easier access to well-qualified consultants, sub-contractors and distributors.

Thus, in accord with agglomeration theory it can be proposed that SMEs in larger and more diversified business environments tend to win a competitive advantage over SMEs in rural areas. Locational conditions also influence the factor side and internal resources of SMEs, that is their manpower, core skills and networks. Likewise, there are grounds for arguing that location also has an effect on the skills of the entrepreneur or manager, his/her work experiences and social networks. In particular, the ability of an entrepreneur to handle the organisation-environment interface as well as social relations in an international context is obviously one crucial aspect of successful internationalisation.

Notwithstanding the arguments emphasising the role of external resources and local milieus in the internationalisation of SMEs, it is obvious that these firms are also first and foremost dependent on their own internal resources and products in their struggle to win market shares in foreign markets. Figure 10.1 summarises various elements in a model as developed by Vatne (1995).

In the following sections of this chapter, empirical analyses inspired by this model will concentrate on the importance of internal resources to exporting and on small firms dependence on their local business environment. The tendency to export is very much a product of which industry a firm belongs to. For this reason only firms from wood-processing and metal industry are included in this study. This is industries with fairly high export-shares and a wide geographical distribution.

In this chapter we will limit our analysis to simple bivariate descriptive statistics and to a test of the significance of observed differences between three groups of firms; committed, experimental and non-exporters. The generalisation to be derived from such statistics is of course limited, but they still give some indication of important processes, which can explain the export behaviour of firms.

Internationalised firms vis-à-vis home market firms

The empirical data were collected in Norway, Finland and Sweden through detailed interviews with 274 independent, locally owned firms in wood-processing and metal industries having 10 to 199 employees.[67] Some basic information on the surveyed firms is described in Table 10.1.

Table 10.1   Nordic SMEs: the data

	Norway	Finland	Sweden	Total
number of firms	93	80	101	274
wood-processing	42%	29%	35%	35%
metal industry	58%	71%	65%	65%
employees: median	24	19	21	21
share of exporters %	58%	45%	60%	55%

Source: own data

Thirty five per cent of the surveyed SMEs are wood-processing firms. The median number of employees was 21. More than half of these SMEs had exports in that year, but only a few had proceeded further in their internationalisation. For this reason, internationalisation refers to exporting in the following empirical analyses.

The SMEs covered are located in four types of region. These regions (big city, industrial city, small town and rural periphery) differ substantially in terms of size, location and structural characteristics. For example, the big cities under consideration have 200-500,000 inhabitants, industrial cities approximately 100,000, small industrial towns approximately 20,000, and rural peripheries have only minor municipal centres. As a whole, these regions (12 in total) form a diverse environmental setting for the analysis of how the structural features of a local operational environment are reflected in patterns of internationalisation.

To split SMEs into those that export and those that serve only home markets is a very crude dichotomization. In practice, the degree of the internationalisation of exporting firms varies greatly. Some firms rely entirely on foreign demand whereas others have only marginal export deliveries based on unsolicited orders. There are also firms among the minor exporters who do not intend to increase their export involvement or launch other international operations.

In the analysis, exporting SMEs are further divided into two groups; committed exporters and experimental exporters. This grouping is a simplified version of the one developed by Cavusgil (1984). A committed exporter has exports amounting to at least 20 per cent of its turnover or exports to at least three countries. Those SMEs with export shares of less than 20 per cent and export to no more than two countries are considered to be experimental exporters. Table 10.2 gives basic information about the SMEs falling into these groups.

Table 10.2   Distribution of the Nordic SMEs (N=274) as committed exporters, experimental exporters and home market firms (non-exporters) by countries and by regions, per cent

		committed exporters	experimental exporters	non-exporters
Norway	N=93	37	26	38
Finland	N=80	31	14	55
Sweden	N=101	50	11	40
wood industry	N=97	34	20	46
metal industry	N=177	43	15	42
exporters	N=155	70	30	
total	N=274	40	17	43

Source: own data

About two-fifths of the SMEs are classified as committed exporters and somewhat less than one-fifth as experimental exporters. Non-exporters,

117 in number, form the biggest group. There are clear-cut differences in the distribution of these three groups by country. The proportion of committed exporters is highest in Sweden, where manufacturing industries have a long-standing tradition of international operations. Because of the geographical and cultural proximity to Sweden, there are relatively low export barriers in the path of Norwegian SMEs seeking to enter Swedish markets, which can be seen in a fairly high proportion of experimental exporters. In contrast, the linguistic and geographical distance of Finland from its Nordic neighbours and the Western European markets has obviously constituted an effective export barrier. More than half of the Finnish SMEs in the present data set is non-exporters. The differences by sector between the three groups of SMEs are not as great as the differences by country.

Some 43 per cent of the total turnover of the committed exporters was sold abroad whereas the respective share was only about seven per cent for the experimental exporters. For obvious reasons, the share of the local sales area (labour market district) was lower for the former group (less than 10 per cent) than for the latter group (about 20 per cent). The sales of non-exporting SMEs were on the average distributed about equally between the local area and the rest of the country. Still differences in profitability between the three groups are small.

Following the argument developed in the previous section, internationalisation can be assumed to demand strong internal resources and/or active search for external resources. Here, the internal resources of the three groups of Nordic SMEs are compared first. To do this we use simple analysis of variance with nominal data chi-square tests. After this, the methods used by SMEs to gain access to external resources are explored.

Internal resources

Clearly, the internal human and monetary resources of a firm depend to a greater extent on its size. Yet the quality of the resources (e.g., the competence of employees) and their compatibility with the business idea of the firm are also of decisive importance. As the earlier research evidence suggests, the key resource of an SME is often the owner/manager to the extent that the continuity and development of its activities rely on him or her. Furthermore, technology and organisation can be conceptualised as resources due to the fact that they influence the efficiency of the utilisation of human and monetary resources.

A summary of differences in the internal resources of the three groups of SMEs under consideration here is presented in Table 10.3. The table reports the mean value for different variables measuring the internal resources of the firm for each of the exporting-groups and also indicates if the differences between the groups are statistically significant.

Table 10.3   Committed exporters, experimental exporters and non-exporters: differences in internal resources in 1991

	committed exporters	experimental exporters	non-exporters	sign. level
size				
- number of employees-median	30	29	16	**
-average	45	39	24	**
turnover, in million FIM				
-median	17	10	6	**
-average	24	16	11	**
internal division of labour				
- number of departments	4.6	4.1	3.2	**
production process				
entrepreneurs' assessments, scale 1-5				
- technological level	2.9	2.7	2.3	**
- degree of automation	2.6	2.2	1.7	**
entrepreneur				
work experience, % of entrepreneurs				
- big firms	55	49	47	
- consulting firms	19	20	13	
- foreign firms	32	29	16	*
- other sectors	66	47	47	**
higher education, % of entrepreneurs				
- at least 4 years	23	21	9	*
- 2-3 years	36	41	36	
- less	40	39	56	*
language skills, % of entrepreneurs				
- English	87	84	66	**
- German	60	44	29	**
employees				
- share of white-collar staff, %	29	22	20	**
change in educ. level, % of SMEs				
- increased	61	52	39	**
education, % of white-collar staff				
- university or equivalent	14	12	13	
- college	37	24	30	
work experience, % of white-collar staff				
- big firms	20	20	21	
- consulting firms	6	3	4	
- foreign firms	10	6	9	
language skills, % of white-collar staff				
- English	58	46	49	
- German	22	12	16	*

** .01; *.05
Statistically significant difference (crosstabulation, χ^2, or oneway variance analysis, t-test)

The firms labelled committed exporters are generally bigger than the experimental exporters and non-exporters in terms both of their turnover and numbers of employees. Their internal division of labour is more specialised, and their technological level and degree of automation are higher.

In addition, they have a higher percentage of white-collar staff, the educational level of employees has increased during recent years, and owners/entrepreneurs are both in terms of educational background, language skills and experience with foreign firms and other sectors more competent than their counterparts in the other two groups under comparison. In contrast, there are no statistically significant differences between these three groups of SMEs in the educational background, work experience and language skills of their white-collar staff. In fact, non-exporters are somewhat stronger in terms of these internal resources than experimental exporters.

In most instances it is the non-exporting group which is significantly different from the two exporting groups. Still, the experimental exporter seems to be positioned in the middle, between committed exporters and non-exporters in regard of access to internal resources.

In sum, these findings lend support to the view that the qualities of a firm's internal resources and its entrepreneur have a major impact on the internationalisation of a SME. However, important reservations must also be expressed. The analyses use only simple bivariat statistics.[68] In particular, it should be emphasised that the above findings are based on cross-sectional data and, as a result, it is difficult to make sweeping generalisations about the extent to which the stronger internal resources of the committed exporters are a cause or simply a consequence of their international operations.

External resources

The previous discussion suggested that there are several possibilities open to an SME in mobilising supportive external resources. Conventional market transactions are used, for instance, in the purchase of business services. In addition to market relations, an SME can have long-term co-operative or network relationships in, among other things, product development, marketing and sub-contracting. It is also possible that business ties and social relations are intertwined in the sense that personal contacts play a major role in the search for information and in maintaining and upgrading competitiveness. In general, gaining access to external resources is a dynamic process in which a firm's internal resources, business environment and networks, and the entrepreneur's social position and problem-solving capacity are linked to one another. Table 10.4 summarises findings from the present study on various means of gaining access to external resources.

As a whole, the data does not support the view that exporting SMEs would be more active than non-exporting SMEs in acquiring external resources. The differences between the three groups of SMEs in their use of external resources are not at all as prominent and systematic as in the levels of their internal resources.

In the data collection process special attention was paid to entrepreneurs' contacts and inter-firm co-operative efforts. Entrepreneurs were asked to assess their professional and social ties and their firm's co-operative arrangements. Since the data collection was based on interviews, the qualities of co-operative relationships (e.g. market versus inter-firm relations) could be evaluated jointly with the respondents.

The differences between the three groups proved to be fairly small in the case of entrepreneurs' social networks. Somewhat more than half of the entrepreneurs reported meeting with their local colleagues at least weekly on professional matters, and somewhat less than half on social matters. With regard to the entrepreneurs' use of their sub-contractors, competitors, customers and other relevant actors as sources of information, the only statistically significant difference was that the experimental exporters are more often in contact with their sub-contractors than are the committed exporters and non-exporters. Although a higher proportion of exporters than non-exporters estimated having to have at least monthly contacts with their competitors, customers and other relevant actors (such as public authorities), these differences fail to be statistically significant. Even so the data seem to indicate that experimental exporters are the group mostly involved in external social relations with the purpose of collecting information of importance for the company.

As far as sub-contracting as a form of co-operation is concerned, the non-exporters make more deliveries to other local firms, and the committed exporters utilise more sub-contracting from non-local sources. This dissimilarity is based on the roles of these enterprises in a value chain. Of the committed exporters, 79 per cent primarily produce their own products[69] whereas the respective proportion in the case of the experimental exporters is 73 per cent and 51 per cent in the case of non-exporters.

Table 10.4 also shows that the more committed a firm is to export, the more formalised are its co-operative arrangements. Put differently, firms with no or few foreign relations organise their inter-firm relations in an informal and often occasional way. Still, the differences between the groups are small and are not statistically significant.

Table 10.4   Committed exporters, experimental exporters and non-exporters: external resources in 1991

	committed exporters	experimental exporters	non-exporters	sign. level
entrepreneur's information network				
contact at least weekly, % of entrepreneurs				
- professional contacts locally	55	60	55	
- social contacts locally	46	39	41	
contact at least monthly, % of entrepreneurs				
- sub-contractors	57	78	57	*
- competitors	32	38	31	
- customers	80	82	70	
- other organisations	41	35	29	
co-operative relations				
- average number	2.2	2.3	2.0	
- location of most important partner				**
local area, % of SMEs	30	47	54	
abroad, % of SMEs	24	3	2	
forms of co-operation, % of SMEs				
- joint ventures	31	30	25	
- contract-based long term co-operation	55	52	47	
- informal long term co-operation	58	61	65	
- occasional informal co-operation	75	80	76	
sub-contracting, % of turnover				
- sub-contracting from the local area	9	13	8	
- sub-contracting from outside the local area	13	6	8	*
- sub-contracting deliveries to the local area	7	12	36	**
- sub-contracting deliveries outside local area	39	33	29	
business services, share of SMEs purchasing only locally, per cent				
- accounting and bookkeeping	84	85	90	
- computing	75	80	76	
- advertising and marketing	65	70	70	
- transport	76	76	79	
- management consulting	49	45	44	
- technological consulting	55	70	61	
- legal services	72	76	76	
- educational services	53	68	68	
- financial services	84	89	94	
- public services to firms	57	64	42	

** .01; *.05
Statistically significant difference (crosstabulation, χ^2, or one-way variance analysis, t-test)

On average, exporting SMEs purchase more business services than their non-exporting counterparts, but this difference is not statistically significant. As is clear in Table 10.4, there are no systematic differences between the committed exporters, experimental exporters and non-exporters in the sense that any one of these groups relies more than the others on the local supply of business services. However, committed exporters seem to be less dependent on the support of external services than are the two other groups.

Even though there were no statistically significant differences between the three groups, the data do indicate that committed exporters are more self-sufficient in business services, more actively involved in non-local inter-firm relations and sub-contracting, and more inclined to formalise such relations. This could partly be explained by the uneven distribution of internal resources between the three groups. Committed exporters are larger, are more involved in routinised large scale production, have more internal capacity to plan and manage their external relations and can more easily handle the problems of rising risk and transaction costs in new markets. Experimental exporters, on the other hand, do not seem to be blessed with the same quality of internal resources and are generally more involved in experimental situations where high risk and trial and error characterise their business activities. Such firms seem to be seeking more information from their external environment, are more involved in occasional co-operation and use external services more intensively than their more established competitors.

The surveyed entrepreneurs were also asked to assess the motives for and forms of their most important co-operative relationship. These assessments are summarised in Table 10.5.

The data suggest that motives for co-operation differ among SMEs in a statistically significant way in two respects. Export promotion, as an accentuated motive for internationalised firms is self-evident. In addition, both groups of exporting firms put stronger emphasis on process development as a motive for co-operation. As regards forms of co-operation, there are more differences between the three groups of SMEs. Those concerning the upgrading of the production process, improvement of products, sub-contracting, information diffusion on technology and delivery arrangements have statistical significance. In general, the two groups of exporting firms place greater value on all forms of co-operation compared with non-exporting firms. It is again interesting to find that the experimental exporters seem in several respects to put more emphasis on co-operative arrangements than do the committed exporters.

Table 10.5   Qualities of the most important co-operative relationship (averages, scale 1-5)

	committed exporters	experimental exporters	non-exporters	Sign. level
Motives				
capacity problems	2.67	2.59	2.70	
economies of scale	2.87	2.63	2.45	
product development	2.86	3.00	2.67	
process development	2.78	2.73	2.24	*
reliability of delivery	3.13	2.95	2.89	
new home market areas	2.20	2.24	2.32	
new home market niches	2.32	2.53	2.17	
export promotion	3.02	3.07	2.07	**
Forms - activities				
visits to the partner	3.48	3.78	3.32	
follow-up of the partner's performance	2.94	3.03	2.61	
upgrading of production process	3.10	3.19	2.50	**
improvement of products	3.55	3.45	3.02	*
sub-contracting	2.46	3.29	2.94	*
joint machinery, personnel	1.85	2.19	1.60	
information diffusion on technology	3.11	3.06	2.61	*
information diff. on market cond. and prices	3.30	3.45	3.16	
delivery arrangements	3.00	2.94	2.11	**
joint marketing	2.82	3.19	2.70	

(two-way variance analysis, t-test; ** .01, * .05)

Committed exporters by country and by region

Analyses of the previous section linked internal and external resource use to the different levels of internationalisation of Nordic SMEs. In this section, the comparison is confined to the resources of the most internationalised SMEs that are committed exporters. Both the national and local operational environment can be assumed to influence their characteristics. From the perspective of Western European markets, these two milieus can be conceptualised as overlapping levels conditioning the activities of internationalised Nordic SMEs.

As far as reasons for country-specific differences in operational environment are concerned, potentially important factors include differences in industrial structures and economic performance. However, the three Nordic countries have a great deal in common in these respects.

They have been open economies for a long time and their production structures have evolved as parts of the Western European economy according to the conditions of international trade. Of the three Nordic countries under consideration here, Norway is most specialised in her exports from the resource-based sectors and Sweden has more features of intra-industry trade than her Nordic neighbours (see, for example, Lundberg 1992). These structural dissimilarities derive from the specific resource endowments and the historical peculiarities of industrialisation in the three countries. They have also created country-specific practices and institutions for international contacts, which are also reflected in the SME sector. With regard to recent macro-economic and industrial performance, the record of these countries is not dynamic. The recession of the early 1990s hit Finland particularly hard, which is obviously reflected in the cross-sectional data of the present analysis.

The regions under consideration here - big city, industrial city, small industrial town and rural periphery - illustrate the internal differentiation of the Nordic countries. The regions in a certain category resemble each other in terms of their structural characteristics and their roles in the regional division of labour, although they are not, of course, identical. These four local milieus - defined in the data collection as labour market areas - can be assumed to provide SMEs and especially their internationalisation with very different preconditions and resources. The relevant differences concern, among other things, distances and transport costs, population density, size of local markets as well as the supply of labour, business services and sub-contracting.

The relative proportion of the committed exporters - as exporters in general - of the SMEs in question (see Table 10.2) is lower in Finland than in Norway and Sweden, but in Finland SMEs export a higher share of their turnover. In Finland the committed exporters sold 53 per cent of their total turnover abroad whereas the respective figure was 42 per cent in Sweden and 39 per cent in Norway. In contrast, inter-regional differences in these shares are fairly minor. The percentage is lowest in the category of small industrial towns (39 per cent) and highest in big cities (50 per cent). In industrial towns it is 40 per cent and in rural peripheries 44 per cent. Not surprisingly, these differences are linked with the roles of firms in a value chain: the share of the SMEs primarily producing their own products is much lower in small towns.

There are only a few significant country-specific and region-specific differences in the internal resources of the committed exporters. The Swedish entrepreneurs evaluate their firms' levels of technology and their educational backgrounds lower than do Norwegian and Finnish entrepreneurs. However, a higher percentage of white-collar staff have a university degree in Sweden and in Finland in comparison to Norway. The only statistically significant region-specific difference in the characteristics of employees concerns changes in their educational level. It has increased in 91 per cent of the committed exporters in small towns, but only in 39 per cent of those in big cities. A tentative explanation is the high degree of restructuring which has recently occurred in the small industrial towns, improving the availability of qualified manpower. In general, the relatively small country- and region-specific differences in internal resources might be due to the increased homogeneity of the competitive pressures facing the committed exporters. There is an important feedback process in operation here: the resources necessary for export activities are also upgraded as a consequence of operations in the increasingly competitive European markets. Table 10.6 gives a summarised comparison of the resource profiles of the committed exporters.

There are more country-specific and region-specific differences in characteristics related to external resources than to internal resources. However, although the resource supply available to the international operations of SMEs can be assumed to vary at least by the categories of region concerned, the differences behind the summary findings presented in Table 10.6 do not seem to be linked in a straightforward way to any hierarchical pattern. As a matter of fact, country-specific factors have a more prominent role than region-specific factors. On the average, the Norwegian committed exporters have more partners in their most important co-operative relationships but non-contract based long-term co-operation is more common in Finland. The structural differences in industrial base and traditions are clear in sub-contracting, which is much more common in Sweden than in Finland and Norway. There are also some country-specific features in the local purchase of business services, but these findings should be qualified by the fact that there are differences in the provision of more sophisticated business services in-house. With the exception of export consultancy, the Swedish committed exporters tend to rely more on in-house services. Finnish SMEs generally have higher frequencies in all kinds of external consulting services for promoting their export. This may be due to the fact that Finnish exporters entered international markets later than exporting SMEs in Norway and Sweden.

As far as region-specific differences are concerned, SMEs in small towns are more inclined to be involved in short-term occasional co-operation. This might be due to their more frequent role as sub-contractors. The differences in the purchase of business services obviously reflect major differences in supply conditions: only a few business services are locally available in rural peripheries and small towns.

Table 10.6   Committed exporters: differences by country and region in 1991

Internal resources	country	region
production process		
entrepreneurs' assessments		
- technological level	**	
entrepreneur		
- technical or commercial education	**	
employees		
- change in educational level		**
education, per cent of white-collar staff		
- university or equivalent	**	
External resources		
co-operative relations		
- number of partners	**	
- other long term co-operation	*	
- occasional co-operation		*
sub-contracting/turnover		
- sub-contracting from the local area	**	
- sub-contracting from outside the local area	*	
- sub-contracting deliveries outside the local area	*	
business services, share of SMEs purchasing only locally		
- transport	*	*
- management consulting		*
- technological consulting	**	
- legal services		**
- public services to firms	**	

** .01; *.05
Statistically significant difference (crosstabulation, χ^2, one-way variance analysis, t-test)
Note: The same internal and external resources as in Tables 10.3 and 10.4 are compared. Only statistically significant differences are included.

The Swedish committed exporters view both their local industrial climate and local public support to industry in the most positive terms. Although the Finnish entrepreneurs in particular have a much more negative view on these issues, their participation in co-operative relations is not essentially weaker. This finding raises the question of whether inter-firm co-operation could in practice have been interpreted in a different way by respondents in these countries.

Concluding remarks

The internationalisation of SMEs is usually confined to exporting, as is the case in the present study. Here, SMEs were divided into three groups, committed exporters, experimental exporters and non-exporters. Export shares and the number of export countries were used as defining criteria. In the first part of the empirical analyses, these three groups of Nordic SMEs were compared in terms of internal resources and potential means of acquiring external resources. In general, committed Nordic exporters were found to be bigger and have in several respects stronger internal resources than the other Nordic SMEs. As far as their vertical and horizontal co-operative relations were concerned, the committed exporters were not very dissimilar from the other two groups under comparison. The most noteworthy exception to this concerned their co-operative partners, which were more often found abroad.

The second part of the empirical analyses dealt with inter-country and inter-regional differences in the resources of the committed exporters. From the perspective of Western Europe, individual Nordic countries and their regions can be interpreted as two overlapping levels of SMEs' operational environments. The four types of region under consideration (big city, industrial city, small town and rural periphery) represent different positions in the hierarchical settlement system of these countries, and they also have clear-cut roles in the geographical core-periphery structure of each country.[70]

The empirical findings on the inter-regional and inter-country differences in the resource characteristics of the committed exporters are fairly unambiguous. There is no straightforward link between the resources and locations of the committed exporters, but the structural pattern is much more complex. In particular, differences by country and by region in internal resources are fairly small, which is presumably at least partly caused by the competitive pressures of the market. In external resources, there are more inter-country differences than inter-regional differences. Differences by country are obviously related to peculiarities in their industrial structures and also to their recent economic performance. Of the former determinants, the relatively advanced sub-contracting systems reflected in parts of the SME sector are a case in point. The significance of the current developments can be seen, for instance, in the emphasis given to restructuring in the assessments of the Finnish entrepreneurs who, at the time of this study, have been struck by a deep recession. In general, the 'country profiles' of the SMEs call for more detailed analysis.

As far as the limitations of the analysis are concerned, it is important to observe that the quantitative description of firms' external resources has focused only on potential sources. Their actual significance is not evaluated here. For instance, it is obvious that a co-operative relationship can be conceptualised in different ways by entrepreneurs and can vary enormously in strategic relevance for different SMEs. Yet an even more important qualification concerns the interpretation of the findings. They do not imply that local or regional milieus are insignificant for SMEs. For instance, the smallness and separateness of localities has probably contributed to differences in industrialisation and also to the fact that in these Nordic countries there are only a few dynamic and specialised new industrial districts based on SMEs.

In any case, the tentative empirical findings raise interesting questions concerning the underlying model of SME internationalisation. Its theoretical background is based on a division of labour and network production in which a number of firms participate in a complex production process. In contrast, the findings of this study suggest that most SMEs produce relatively simple products with less need for many partners. This has an effect on their need for and ways of acquiring resources. It is also clear that this kind of product palette and its related role in the division of labour are not independent of the spatial conditions of production in the Nordic countries. Especially those Nordic SMEs, which are located in peripheral regions and small communities, must take for granted that the supply of local supportive resources is relatively scarce. This is a special problem if exogenous factors such as internationalisation change the conditions of competition after the establishment of a particular firm. Those firms which are for some reason located in peripheral regions and small communities have to utilise different ways of gaining access to strategic resources for their internationalisation, relying more on internal resources and contact potential. This setting is, however, offset by the fact that the traditional sectors of the periphery, wood-processing being the representative example, have inherited well-established international contacts. These resources - the resources of the relevant national industrial cluster - complement or might even compensate for local resources. Here, the resources of a local milieu and a national industrial system are in practice intertwined in the internationalisation of SMEs.

11 Decline and Renewal in Industrial Districts: Exit Strategies of SMEs in Consumer Goods Industrial Districts of Germany

Eike W. Schamp

Introduction: entrepreneurs in declining districts

Spatial concentrations of SMEs linked into specific production chains, and located in non-urban environments, have increasingly been analysed using the concept of the industrial district. It is tempting to speak either of an increasing 'stylisation' of facts on an ideal type of industrial district or of the concept now becoming a buzzword in scholarly debate. In fact, the concept of the industrial district emphasises the specific network character of interrelationships between enterprises and local institutions. In addition to inter-firm linkages in a given local production chain, non-market exchanges of goods, information and people create further external economies for firms. These are backed by "a common cultural and social background linking economic agents and creating a behavioural code and a network of public and private institutions supporting the economic agents acting within the cluster" as Rabelotti (1995), among many authors on industrial districts, put it. Thus, external economies are socially constructed and maintained; they are 'embedded' (see Oinas 1995 for a more cautious interpretation of this concept). Many studies consider the district as a harmonious *ensemble* of small and medium sized enterprises, and consequently do not analyse entrepreneurs and entrepreneurial behaviour in the district very deeply. The district is more or less conceived of as a given collective organisation of production which creates a stable 'contextual logic' for entrepreneurship, based both on the entrepreneur business and personal networks (see for example, Johannisson et al 1994).

However, through time, industrial districts are only as stable as their forms of societal organisation. Increasing evidence emphasises contingent factors, which are responsible both for the emergence, growth and success of a district and for its decline. As Bianchi (1998) put it, "industrial districts are concrete geo-historical formations and not abstract, timeless constructions. Thus, it is legitimate to study the processes of genesis, decline and transformation" (p.97). It is, however, not surprising that the majority of district studies look primarily at success stories, and hence more on the genesis, growth and maturity of districts. Studies on declining industrial districts are very few.

A further question is whether the industrial district concept can be applied to all spatial concentrations of SMEs in a given local production chain. Most spatial clusters of SMEs in a local production chain currently do not demonstrate characteristics of the concept in its purest sense. This is particularly the case for traditional consumer goods industries. Here forward and backward linkages may be scarce, a common cultural background may no longer exist, and the only characteristics that apply may be a common history and the establishment (sometimes recent) of public institutions to support SMEs in the region. As a result, some authors have contested the narrow definition of an industrial district and have widened the term to a number of very different spatial clusters of industries (among others, Horvers and Wever 1989, Markusen 1996, Park 1996). The case of 'old industrial districts' may be different again. Although they currently lack many characteristics of a 'working' district - and they may never have been textbook models of districts - these old districts still enjoy assets external to the individual enterprise stemming from a common history, common use of supporting institutions and the like.

There is good reason to assume that the contextual logic for entrepreneurial behaviour in declining districts is distinct from what has been described for 'working' industrial districts. Empirical evidence, however, is scarce (see for example, Dunford et al 1993, Amin and Thrift 1992). In his recent study on the textile and clothing industrial district of Reutlingen, Staber (1997) revealed two interesting facts: first, there was almost no evidence for the existence of any kind of networks between enterprises; and second, vertically integrated enterprises had a huge survival advantage compared to specialised enterprises (Staber 1997). This, however, runs completely against current concepts of 'flexible organisation', 'new networks' and 'embeddedness' in districts. Firms obviously follow different trajectories resulting in the fragmentation of the district and the isolation of firms.

This suggests that the strategies of entrepreneurs in old districts need to be looked at more closely. One may question the notion of (long-term) strategy in the case of owner entrepreneurs as their ability to control their firms' environments seem to be rather restricted. Instead, it may be necessary to emphasise the 'alert firm' (see Chapter 12). This, however, seems to be a concept focusing too strongly on the reactive capacities of an enterprise, not sufficiently taking into account the long-term 'leitmotiv' owner entrepreneurs generally have – for example, a sufficient level of income for the entrepreneur's family in a context of self-determination. We can agree that many SME entrepreneurs do not have a proper strategy. They are stuck with what they have done for a long time, act defensively when their market environment is changing, and do not dispose of, do not use, or do not create assets, either internally nor externally. These SMEs mainly rely on their long-term experience in processes and products. Hence, they rely on tacit knowledge. They seem to be locked-in to the traditional trajectory of a given industrial district. Their learning capability seems to be rather restricted (Asheim 1997). These firms form the traditional declining part of an old industrial district, and represent its 'sclerosis'. The survival of these firms is uncertain. Exit via bankruptcy and closure is probable.

It is argued in this chapter that there are, however, innovative owner entrepreneurs in some old industrial districts who follow an explicit strategy in choice of products and markets, in making new uses of the external assets of districts, and in creating new assets that enable them to contribute to the renewal of the district as a collective organisation. There is a long tradition in research on SMEs and entrepreneurs, particularly in older development theory, that emphasises the role of the 'foreigner' as entrepreneur in the change of given societal contexts. However, 'foreigners' may only create their own networks and may largely remain outsiders in the local society. In other words, they lack 'local embeddedness'. Furthermore, the entry of new entrepreneurs into declining old districts seems to be rather unlikely. So entrepreneurs *combining* both the role of a 'foreigner' and yet being more or less locally embedded are needed to revitalise old industrial districts. Here, the question of a new generation of local entrepreneurs comes to the fore, for example, entrepreneurs' children with experience from 'abroad' coming back into the district. 'Exit' under these circumstances then means the survival of the firm as a legal entity but change in its activities, its position in the traditional production chain and even in its sector of operations. If these firms succeed, 'learning' in the district seems to depend more on non-local experience than on some kind of collectivity in the district. It can be suggested that this holds true for both incremental and more radical innovations. Hence, it is necessary to question a too narrowly defined concept of 'local learning' (but see Asheim 1997).

What will be discussed in this chapter, is not a mere problem of the ageing of an industrial district. Given the societal change in the organisation of production and consumption that has been taking place since the 1970s, surviving firms have to adjust to different and changing economic environments. That means, for example, that firms can no longer hope to find mass markets and use traditional forms of distribution. Instead, they have to define market niches, upgrade product quality, introduce new products and new forms of distribution, and reorganise production: in short, they have to 'learn'. Proximity, then, seems more relevant for imitation by these new entrepreneurs than for some kind of collective learning (see, for example, Maskell et al 1998). And, furthermore, something more radical than incremental learning is required (see for example, Asheim 1997 and Chapter 8 in this volume).

An old industrial district is, therefore, characterised by a peculiar mixture of different kinds of survival, exit and entry. Recently, Clark and Wrigley (1997) have reconsidered the problem of firm exit in industrial geography using the concept of exit sunk costs. Although they focus on the strategies of large firms, the same strategies of exit to ensure the survival of the firm seem open for SMEs as well: (strategic) reallocation, restructuring and change of corporate form. There is an uneven distribution of competencies among firms in an industrial district and, hence, an uneven distribution of capability to change strategies. A somewhat traditional way of rejuvenating industrial districts has been to change corporate form. Bankruptcy, for example, which has occurred more often in declining districts, may not be a challenge but a chance for the district if new entrepreneurs take over and restructure those firms. It is, however, more likely that new entrepreneurs are less prepared to invest into old industrial sites and production lines. Strategic reallocation is strongly bound to the existing assets and competencies of the firm, and the capabilities and willingness of the entrepreneur to change strategically. Strategic reallocation is more likely for larger, more integrated firms. These are opportunities for firms to break lock-ins in traditional ways of thinking using and making which characterise old industrial districts. Small scale firms in traditional sectors mainly rely on their experience in processes and products, and hence on tacit knowledge. As tacit knowledge is bound to individuals, it is not surprising that major technological changes in sectors can only be expected from outside, through either a new generation of entrepreneurs or non-local entrepreneurs. But, what is the likelihood that these entrepreneurs will enter a declining sector?

The main argument of this chapter is that entrepreneurial strategies are strongly bound to the district as a special production environment through its assets that are external to the firm (Figure 11.1). This traditional, established environment creates barriers to more fundamental changes in firm strategy and, hence, produces certain lock-in effects. But, simultaneously that environment can be used to build up new external assets and to pursue new strategies that finally result in a new kind of 'exit' from the old district.

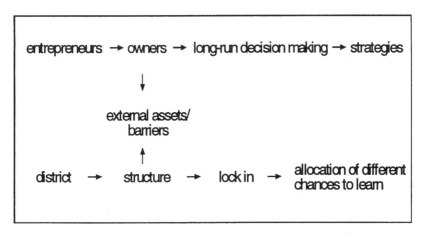

Figure 11.1   The district as framework for entrepreneurial strategies

Most empirical work on industrial districts has been done on either high-tech new industrial districts, where mostly intermediate products are produced, or design-intensive older districts engaged in consumer good production. This chapter focuses on consumer goods production in spatial SME clusters in Germany. These are low-tech enterprises that possess a limited array of competitive strategies and, furthermore, many of them are still not design intensive. Small and medium sized firms in essentially traditional consumer goods industries are often spatially concentrated in Germany. Porter (1990) gives the example of the cutlery industry in Solingen, but there are many more localised production systems; for example, leather goods production in Offenbach (Bertram, forthcoming), household textiles in Bocholt (North-Rhine-Westphalia) and Laichingen (Baden-Württemberg, see Behlke 1995), customer jewellery in Neugablonz

(Southern Bavaria, see Rehle 1996), and wicker ware production in Lichtenfels/Michelau and footwear industries in the Pirmasens areas. It is on these last two 'districts' that this chapter will concentrate.

The Lichtenfels/Michelau and Pirmasens districts: common traits

This chapter adopts Bianchi's (1998) approach by taking a long-term perspective on two spatial clusters of SMEs, which might have had, in former times, some characteristics of an industrial district. It discusses two different spatial clusters of SMEs in decline and transformation, both located in rural areas in border regions. These are the cluster of crafts in wicker work around Lichtenfels/Michelau in Upper Franconia, on the northern border of the state of Bavaria, and the cluster of footwear production in the Pirmasens area, on the French border in southern Palatinate (Figure 11.1). In both areas, neither wickerwork nor footwear production are the most important economic sectors today. But long ago, they used to be the economic base of both areas. Upper Franconia is an industrialised rural area, well equipped with a network of small towns and, at the same time, attractive for tourists. It is still an area of relatively low wages and high female labour market participation.

Although both clusters have histories of their own, they share common traits in the phase of formation. First, their origins stem from proto-industrial times, during the 18th century, when territorial rulers fostered the introduction of low wage industrial homework to cope with rural poverty. In Upper Franconia, the bishop of Bamberg created a guild at Michelau in 1770 to make peasant basket makers more competitive and to enable the emergence of crafts. In Pirmasens, a prince of Hesse relocated his residence to this remote place in 1719 to build up a regular army. He initiated felt shoe production to give labour to the soldier's families. This proved all the more necessary when his successor relocated his residence to the city of Darmstadt and abolished the regular army. The origin of both districts was, therefore, based on low technological skills.

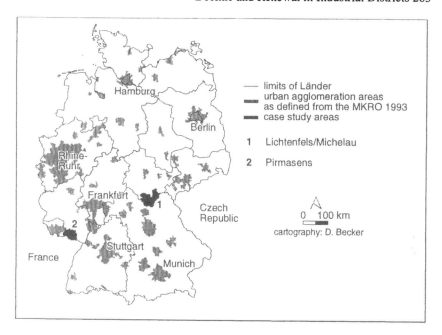

Figure 11.2   The location of Lichtenfels/Michelau and Pirmasens

Second, major growth took place more than a century later, in the period of rapid industrialisation and urbanisation of Germany at the end of the 19th and beginning of the 20th centuries. Both clusters produced cheap low-tech mass products for poorer urban households, and faced enormous growth in demand during this time. Additionally, the wickerwork cluster was a major producer of packaging materials for most of the manufacturing sectors. Third, a new social division of labour emerged during the 19th century, separating in particular different stages of the production chain in craft production, on the one hand, and distributors, on the other. Production was also partly reorganised into factories. Fourth, towards the end of the 19th century, new products were introduced such as wicker furniture in Lichtenfels/Michelau district (Benecke 1921) and leather shoes (partly imitating American models) in the Pirmasens district. This resulted in further growth and investment in supporting infrastructure such as public vocational schools, which were established to upgrade skills in the technologies required. The German *Korbfachschule* (vocational school of wickerwork) was established in 1904 in Michelau, the German *Schuhfachschule* (vocational school of footwear production) in 1927 in Pirmasens. At that time, in the early 20th century, both clusters may have been industrial districts in the proper sense, as existed elsewhere in Germany. There were different production clusters in basket making and footwear production in rural and urban areas, respectively.

After World War I, both sectors were protected from change by autarkic policies in Germany. They even played a certain role during World War II. When Germany formed a part of the liberalised world economy and the wealth of urban households grew in an unprecedented manner after 1950, both sectors developed in a different way. Basket making declined first, from the early 1950s on, partly due to rising wages in Germany and rising imports from abroad, and partly due to competition from the emerging plastics industries. This happened all over West Germany, but as the Lichtenfels/Michelau district had been the main core in this sector, some firms survive until today.

Footwear production started to decline in the late 1960s, when large urban firms relocated production first to remote rural areas such as Pirmasens and, then to neighbouring lower wage countries such as Alsace/France, Italy and Spain.

Thus, the decline of production in both sectors was a result of the collapse in demand for traditional products and the substitution of traditional products by both cheap imports and products from new materials such as plastics. Production retreated to one spatial cluster in each sector, to Lichtenfels/Michelau and the Pirmasens area, in wicker ware and footwear respectively. It would be misleading, however, to conclude that both clusters have faced decline because of the emergence of a more competitive Fordist production in the Fordist division of labour. On the contrary, the footwear district of Pirmasens saw its economic peak in the Fordist era when large shoe manufacturers entered the region with branch plants in search of disposable low cost, usually female labour. Thus, what is described in this chapter might be seen as part of a complex story of the amalgamation of large scale and small-scale production in times of Fordism and post-Fordist change. The point to be made here is that both clusters continued to grow through mass production for low-income mass markets, based on low technologies and low paid labour in remote areas of Germany. During the last three decades, both districts have suffered a tremendous decline in their markets, decline in numbers of firms and employment, and the undermining of local institutions.

The competitive squeeze on firms of both clusters rose incrementally, however, and firms only slowly realised the necessity to shift strategies. Possibilities for new strategies were different, according to the organisation of the districts and the technologies involved - either craft organisation in a limited production chain, as in the wicker industry, or factory organisation on an extended production chain as in footwear production. With the gradual decline of the consumer goods sectors in a vertically integrated industrial district, the complex web of inter-relationships which creates and maintains local external economies was threatened at different stages of the production chain, simultaneously. Firm closures and other forms of exit in different, vertically linked sectors of the clusters contributed to the decline of an industrial district. Surviving firms had to cope with two problems at the same time: first, to reassess their own resources or to create new ones to cater for new products in new markets; and, second, to maintain mutually accessible local external assets.

It is on the most recent strategies of surviving firms that this chapter focuses. An attempt is made to answer two questions. First, what are the individual exit strategies of firms in their struggle for survival and growth? Second, how do actors' different strategies contribute to the preservation of a district through modernisation? It is in this field that the state plays a role. The chapter draws on two field surveys to answer these questions, one in 1996 in the Lichtenfels/Michelau area and the other in 1998 in the Pirmasens area.

Wickerwork in the Lichtenfels/Michelau district

Wickerwork is a typical low wage, cottage industry, which had turned to factory production at the beginning of the 20th century. The technical production chain is simple, as the social division of labour is rather shallow (Figure 11.3). Traditionally, the production chain is: first, trade in raw materials: second, production of wickerwork in cottage industries or by homeworkers and production of wicker furniture in craft shops and small industrial plants; and third, distribution by itinerant craftsmen, through producers' small local shops, or through retail enterprises. The latter mostly applies to distribution from the larger producers of wicker furniture and firms importing of wicker ware. This production chain is supported by local institutions such as the vocational school, the recently created basket fair and the innovation centre. The sector is characterised by family structures, both among entrepreneurs and homeworkers. Wickerwork has always been a handcraft profession that resulted largely in a craft organisation of production or in literal 'manufacturing' with distributors organising the production of homeworkers. In fact, technical skills are low and are mostly bound to on-the-job experience, i.e. they rely on essentially 'tacit' knowledge. The vocational school for wickerwork plays a major role in qualifying 'middle management' in wickerwork industry: i.e. providing basic knowledge for a master craftsman qualification and for some skilled workers in small scale enterprises. Legally, wickerwork is a craft in

Germany. That means that founders of a small-scale enterprise in the sector need to be 'master craftsman' or, at least have to employ a 'master craftsman'. Most male and female homeworkers, however, have only learnt on-the-job.

Although the wicker industry has long since lost its importance for the regional labour market and regional prosperity in the Lichtenfels/Michelau district, it still remains the major spatial cluster of the wicker industry and wicker trade in Germany. Employment in the basket making sector and basket trade has decline tremendously since its peak in the early 20th century. At that time, more than 20,000 people worked in the sector, mostly as homeworkers, and a large part of production was exported to Western Europe and the USA (Benecke 1921, p.32 and 73). Although decline mainly happened through the 1950s and 1960s, it has continued through the 1980s and 1990s (Figure 11.4).

Our survey in 1996 identified 44 surviving firms in the wicker sector, most of them were very small with 10 or fewer employees. From this small group of survivors it is again clear that most producers had failed to cope with declining demand and rising competition from abroad from the 1950s on. Roughly 600 surviving workplaces including some 70 homeworkers could be identified from this survey.

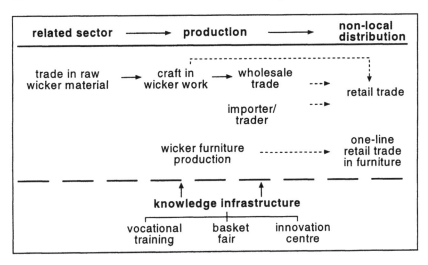

Figure 11.3   The wicker industry production chain

Firm strategies in the wicker industries

The survey did not encompass those surviving firms, which had been successful in changing sector totally. There is sufficient evidence to claim that exit strategies of firms by radically altering production were particularly successful in earlier times. These firms made innovative use of their knowledge on the technical processes in wickerwork and wicker furniture production. Wicker furniture producers who had incorporated carpentry into the firm - which forms the stabilising backbone of most wicker furniture - were able to shift into the upholstery sector. Because many firms followed this trajectory, a new spatial cluster of upholstery businesses emerged during the 1960s and 1970s in the Coburg region, just north of the research area (Haffert 1957). As most upholstery is as low-tech and low skilled as the wicker industry, today this sector and region face the same competitive squeeze from imports. Some wickerwork producers switched into baby carriage production, in the early years of the 'baby boom' after World War II. These firms were later able to use new raw materials, particularly synthetics. Some of them were very successful as they entered the growing sectors of the plastics industries, making products for new markets such as plastic parts for the white goods industry, construction industry or automobile industry (see Schamp 1996, 1997). For example, by far the largest industrial firm in Michelau, with more than 1,000 employees, began by trading in wicker materials, and is now an important supplier in plastic parts to the automobile sector. Firms such as this have left the 'district' as a collective organisation. They no longer participate in the production of externalities in the wicker industry.

In the wicker industry production chain, 26 firms agreed to in-depth personal interviews, among them 15 producers, 8 distributors/ importers and 3 traders in raw materials. From these interviews a district-type model of the local production system among firms which stayed broadly in the wicker industry can be re-constructed. Furthermore, they allow different types of firms to be identified according to their 'exit' strategy from traditional production. These firms can be differentiated both by the degree of radical change in technology and products they have pursued and by their ability actively to alter their institutional environment, i.e. to 'create' the locational factors they require.

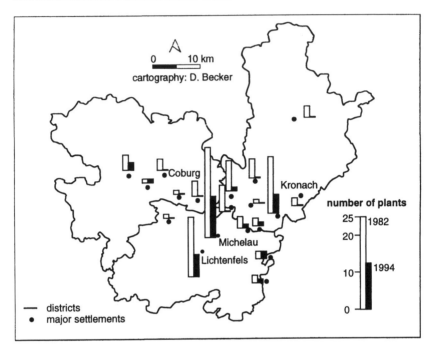

Figure 11.4   Spatial retreat of the wicker industry in Upper Franconia, 1982 to 1994 (Förth 1995)

The most successful firm in the wicker industry, a genuine 'innovator', is Flechtatelier Schütz. Although having grown to be medium sized industrial company producing high quality, well-designed wicker furniture (lounge, upholstered and bedroom furniture) it still uses a craftsman-like title 'Atelier', thus signalling the quality and individuality of a craft shop. More than 250 employees are employed in customised production. This is a success story combining local origin with foreign experience. Kurtz Schütz came from the region, had learnt wicker work at the vocational school in Lichtenfels, had designed models of wicker work, had been a teacher at that school, and had emigrated to Switzerland in the 1950s where he learned rattan furniture production. Schütz came back to Lichtenfels in 1960 and started producing different wicker work and using new raw materials because of a lack of capital to start producing rattan furniture. During the 1970s, however, he shifted to furniture production. Because he managed to combine different innovations in production, technology and distribution, and was able to bring together diversified expertise, the firm grew dramatically during the 1980s and 1990s.

First, Schütz was able to make use of workers who were well trained at the vocational school. "I was lucky to find many very well trained staff at the vocational school in Lichtenfels who became excellent skilled workers. They were, for example, able to read and translate a drawing" (personal interview). According to Schütz, these workers still form the backbone of the workforce today. Second, Schütz increasingly looked for new ideas in design from others, although he still prefers to design himself. At first, independent designers were asked, then his daughter studied design at the vocational school. His son was trained in business science and is responsible for business management today. Third, and most import, was a new distribution strategy since, in the beginning, one-line retail shops were rather reluctant to sell his products. Schütz organised a truck as a rolling show room and sent it to major events in large cities but did not, and still does not, sell directly to private clients. In doing so, he got direct contact with the final customer and at the same time gained the attention of one-line retail shops. This enabled the company to stay in the high priced segment of the market where cost competition is lower. Nowadays, the company has a well developed distribution system through selected furniture retail shops in Germany and in neighbouring countries such as the Netherlands, Austria and Switzerland. Fourth, Schütz made use of the annual 'Basket Fair' at Lichtenfels to bring visitors to his show rooms, presenting current and future models and testing (future) client's tastes. In fact, Schütz's company was successful in both combining its own competencies in process technologies with new competencies in design and distribution and using local institution such as the vocational school and the fair as resources to enhance competitiveness.

Finally, Schütz, has been able to establish an extended personal network by taking over new commitments in the guild of wicker craftsmen. In this way, he has been successful in creating a new local institutional environment for his firm. He became president of the wicker craftsmen guild in Germany (*Bundesinnungsmeister*) and started to use contacts with politicians to improve the qualification base of the wickerwork industry. As a result, Flechtatelier Schütz stopped vocational training at his factory and only took school leavers from the public vocational school, apart from employing many unskilled female workers. In accord with the guild, the school recently changed the teaching curriculum from basket making to wicker furniture. Furthermore, as the guild's president, Schütz was successful in 1995 in attracting public investment from the Land government of Bavaria into a new innovation centre (*Innovations-Centrum des Deutschen Flechthandwerks Lichtenfels*) aimed at technical improvement in wicker furniture making. In fact, this is the only German centre for research into new production technologies and new design forms in wickerwork, both in furniture production and construction. Although both the school's and innovation centre's clientele is not local but national, Schütz was able to construct proximity advantages for his firm and for some other local furniture producers.

Apart from smaller craft shops that were able to follow Schütz's strategy in producing design-intensive wicker furniture - a very small niche market in Germany - more SMEs took the path of becoming 'modernisers'. These firms were not as innovative as Schütz and others, but extended from basket made into new wicker products, such as wicker furniture or lampshades, and also making use of new forms of distribution at the same time. Higher priced lampshades are sold to mail order firms or through handyman retail shops and furniture retailers. Wicker furniture is sold partly through the one-line retail trade, and partly directly to private households. As these firms are increasingly vulnerable to competitive imports, they can make some profits from the initiatives creating a new local institutional support structure.

It comes as no surprise that basket producers and basket traders could not really rely on these new initiatives of their guild master. Wicker craft shops still rely on traditional experience, drawn both from the vocational school and from in-house vocational training. However, they have had to gradually reduce in-house vocational training due to its costs. Only one or two wickerwork craft shops in Lichtenfels improved product quality and established their own retail shops. Most small craft shops still make use of traditional, local, tacit knowledge in production and local distribution channels. These are the second or third generation craft operations, tied to traditional basket making in family-based production, sometimes producing niche products such as basket miniatures. As most of the craftsmen of this type are over 50, it is obviously a sector nearing its end. These are the sector's traditional producers.

Survival opportunities come from external changes in local environments. In search of structural adjustment in the local economy, local authorities have discovered tourism in recent years. In 1978, the town council of Lichtenfels established an annual Basket Fair, which later became well known. The number of exhibitors steadily rose from 20 at the start to more than 50 in 1996. Most exhibitors came from the region but some were from eastern and southern parts of Germany. The town succeeded, furthermore, in getting the official 'brand' of *'Deutsche Korbstadt Lichtenfels'* (Lichtenfels, German basket town). Finally, the municipal basket museum at nearby Michelau, which had been established in 1934 (Dippold 1994), was professionalised by the appointment of an academic director in 1992. Thus, for most of the crafts in basket making, the annual Basket Fair has become a major distribution channel - urban tourists being their major clientele. Crafts in basket making are not only largely 'embedded' in the character of the region as a tourist area, but are also dependent upon it. Few craftsmen follow their private clientele into the large urban agglomerations and only sell occasionally through events in the large shopping centres. Thus, whereas Schütz had constructed his own proximity, these craft firms make use of an established one, building local folklore, which is useful to attract tourists from large cities.

In addition to producers in the wicker industry, a larger sector of wickerwork traders has emerged. As the wicker sector declined, many families owned craft firms shifted from production to importation and trade. They mostly import low cost wicker work from abroad, e.g. from China, and make no attempt to use the local 'brand' for market segmentation. Occasionally, however, traders use local homeworkers for final processes such as sewing basket linings or varnishing. In doing so, they make use of the surviving non-formal vocational training of the traditional craft shops. There is no reason for traders to stay in the region as they have largely left the 'district' sector. Their location can be explained by the inertia of family owned firms.

This heterogeneity of surviving firms shows why the local wicker production system has never really been moulded into an ideal-type industrial district. During times of mass production, competition between firms and homeworkers prevailed. Through increasing fragmentation of the local production chain, there was almost no chance of establishing the social integration of the district, i.e. horizontal or vertical co-operation and trust. In fact, both in earlier studies and the present survey, entrepreneurs expressed their mistrust of others, and particularly of institutions such as the national guild. The local production system of the wicker industry is obviously exposed to increasing fragmentation by the simultaneous processes of decline and renewal.

Footwear production in the Primasens district

Footwear production has been identified as the base of ideal-type industrial districts by many researchers. Many local clusters in footwear production have been studied across the world, in European countries such as Italy, France and Spain (e.g. Scott 1988b, Courault and Romani 1992), or in LDCs such as Mexico, Peru, Brazil or India (e.g. Rabelotti 1995, summarising Nadvi and Schmitz 1994).

The footwear production system seems to be very different to that of the wicker industry, as it involves a far more sophisticated local production chain. In its final stages, the footwear industry is very much an assembly industry, comparable to car manufacturing. There is, in fact, a diversity of suppliers, related firms and supporting institutions in the production chain (Figure 11.5). Over more than a century, a well developed local footwear production system emerged that came to dominate the local economy in Pirmasens and the surrounding district. Firms and plants were small and family owned despite the relocation of production from urban agglomeration areas by larger firms. The larger firms principally produced footwear in the lower and middle price segments of the market, mostly for females; i.e. mass production using cheap female labour and homeworkers, who mostly had learnt 'on-the-job'.

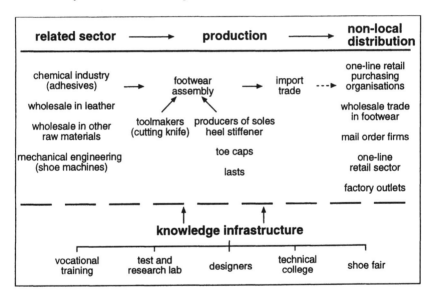

Figure 11.5   The footwear industry production chain

We are fortunate to have a 1970s study on footwear production in Pirmasens, which can be re-evaluated in the light of the district debate (Kuhligk 1976). At the beginning of the 1970s, the district had three distinctive characteristics. First, it had extended a division of labour among specialised small firms, both in parts production and assembly. Flexibility was high as a result of both vertical and horizontal co-operation, for example, when bottlenecks in capacity occurred. Flexibility was further achieved through the use of a significant number of female homeworkers who specialised in specific tasks. The closing down and reopening of firms was a prime engine of labour mobility in the sector. Second, the district had proximity advantages in relation to a host of supporting sectors. An important mechanical engineering sector had emerged in the district. Producers took advantage of both modern machinery and urgent repair services. Local chemical plants and wholesale firms provided special adhesives as well as leather and other parts. There was also local production of lasts and cutting knives (see Figure 11.5), and independent designers created new collections and produced prototypes. Third, public institutions contributed to the production system. A vocational school supplied training and a public research laboratory, founded in 1957 for the development of new materials, new processes and testing, supported production.

Clearly, footwear production in the Pirmasens area in the 1970s revealed many textbook characteristics of an industrial district. There were, however, many signs of ageing and future decline, and some very significant lock-ins were to be found in the region. First, most footwear assemblers had only weak or almost no access to the market. Increasingly, the large purchasing organisations of the one-line retail sector began establishing their own brands, developing their own collections and imposing technical change on the footwear assemblers. Thus, though many footwear assemblers could reduce market risk by supplying these organisations, they also lost market transparency and became dependent on those buyers. As a result, export figures were extremely low, i.e. 5 per cent to 6 per cent of sales. The nearby urban agglomerations such as the Ruhr area and the Rhine-Main area were the prime sales regions. When, finally, wages rose during the 1960s, even in this remote area of Germany, the retail buying organisations began to shift procurement to foreign countries and reduced or even stopped procurement from local footwear producers.

Second, local entrepreneurs in the footwear sector had been able for a while during the 1950s and 1960s to take possession of most local resources and to prevent the rise of new, competing sectors. They kept industrial sites and increased wages to the low skilled workforce but did not fundamentally alter their way of production. When demand fell and supply shifted to imports in the crises of the 1970s, the result was bankruptcy and rising unemployment in the region. A rough estimate puts the number of footwear firm closures in Pirmasens since 1975 at more than 100 (for fuller discussions on firm demography see Bertram and Schamp 1998).

Third, during the subsequent long phase of steady decline in the sector in the Pirmasens area, many small firms attempted to cope with cost competition by reducing the costs of training. They largely shifted the responsibility for training to the vocational school. However, young people increasingly were no longer prepared to enter a declining sector and, consequently, the vocational school saw a drastic decline in first year students. There is now a lack of qualified personnel at a time when the only way out of crisis is to raise the quality of production. Differently to the wicker industry, various attempts to adapt the vocational training system to new technological requirements failed in the footwear industry.

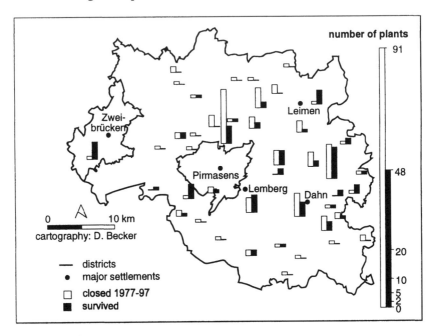

Figure 11.6   Spatial retreat in the footwear industry in the Pirmasens area, 1982 to 1997 (Bertram/Schamp 1998)

As a result, the footwear sector has faced a steady but sharp decline in the last two decades from its peak of activity in the 1960s. At that time, large footwear producers such as Salamander began to relocate production from the urban agglomerations to the Pirmasens area, thus heralding a similar process of spatial concentration into a 'refuge' area as in the wicker industry. These large firms, however, have further relocated production abroad. So, there is now a further spatial concentration of footwear production in the region (Figure 11.6). There are no detailed official statistics on the quantitative effects of the decline on the local labour market. Employment in the core sectors of the production chain, i.e. parts production and assembly in the footwear industry, declined in the Pirmasens area from some 27,000 persons in 1961, to 4,700 in 1997. Employment figures, however, are only a proxy for the decline of the production system, given productivity gains in recent decades. The sector is still contracting in Germany according to recent information. The number of footwear firms in Germany was 288 in 1987 and 160 in 1997. In the same decade footwear production fell from 78.5 million pairs to 40.4 million pairs (Hofmann 1998).

A recent view of change in the footwear industry in the Pirmasens area is provided by a 1998 survey of surviving firms in the different sectors of the production chain using a newly created data base in the absence of detailed statistics on the production chain in the region. This database includes 120 firms in footwear and parts production and further firms in related sectors. Personal interviews were conducted with 40 firms in different sectors of the production system.

The survey did not cover those surviving firms, which had left the footwear production system and entered another sector, as in the case of the wicker industry survey. Many firms were able to shift to other sectors. Larger parts suppliers of items such as toe caps, heel stiffeners and sole products started very early to join the plastics sector. Chemical plants looked for new markets for adhesives or entered the plastics sector as well. The most famous success story in the region was the Kömmerling group, which started in adhesives production for the footwear sector but which became the European market leader in the production of plastic window frame profiles. Machinery firms looked for other clients, such as car manufacturers, or turned from the production of new footwear machinery to trade in used machinery, profiting from the high number of closures in footwear production. In short, the vertical structure of the footwear production system was eroded.

As far as the forces of the erosion are concerned, however, the decisions of larger firms and the purchasing organisations of the one-line retail trade *not* to alter product range and technology but to shift location have been far more important. These firms stayed in the highly competitive lower and middle priced segments of the market but increasingly relocated production to foreign countries. They have subsequently become traders and importers in the Pirmasens area who organise production abroad and distribute in Germany. They sell own-brands, and design and prototype their new collections in the Pirmasens area. Although these firms have largely left the region's production system, they still rely on its traditional knowledge base. This is based on the local availability of designers who have learnt their job through learning-by-doing in footwear production, in design offices or in the public vocational school. Additionally, the public research laboratory still quality tests materials and final products. Thus, the firms still make use of the residual external effects of the old district. The question arises as to how long the local knowledge base can be maintained if production is increasingly disappearing from the region.

While traders and importers play a much larger role in the development of the footwear district, compared to the wicker industry in Lichtenfels/Michelau, 'niche specialists' are of comparatively minor importance. Small family firms have reduced their workforce, some firms to the family alone, using their traditional know-how in production to produce 'traditional' and niche products, such as old-fashioned models or carnival dancing models. As they no longer change collections annually, there is no necessity to buy expensive tools (such as cutting knives). Production is integrated as far as possible. Labour intensive upper production and assembly are done in the workshops. The most serious problem is access to clients.

These traditionalists only sell directly from the workshop. Once more, as happened in the Lichtenfels/Michelau area, these firms profited from the recent public policy to diversify the regional economy and create jobs by fostering the tourism sector in this remote mountain area. The most obvious example is the village of Hauenstein; formerly Germany's largest village engaged in footwear production. It has been seriously hit by firm closures in recent decades, which has reduced the number of local workplaces nearly to zero. The village has recently become the centre of 'shoe tourism', tourists coming from Alsace/France and, during the weekends, from neighbouring large urban agglomerations. A footwear production museum has been opened in one of the old production sites. Surviving small-scale footwear factories have opened sales offices, and a new factory outlet centre only for shoes has recently been established.

Medium sized footwear producers have not had the opportunity to employ such a survival strategy. Many of them have disappeared. Surviving firms are daily in danger of going bankrupt. These 'traditionalists' are rather passively exposed to rising competition from abroad as they stay locked-in with no brand, no direct access to clients and dependency on powerful purchasing organisations. This is partly a generational problem amongst local entrepreneurs. Some of them have no exit strategy. Most of them have relocated production of uppers to foreign countries, particularly to Hungary. To cut costs they sell only 'weak' brands in low and middle priced market segments and are obliged to sell to the powerful purchasing organisations of the one-line retail trade, wholesale traders or mail order firms. Firms and entrepreneurs seem to be at the end of their natural life cycle as they make only weak efforts to change, and they are particularly vulnerable to cost competition from imports.

However, the survey also found a number of 'modernisers' in the footwear sector - family firms looking for survival through the modernisation of products, processes and distribution. Mostly firms have changed strategies with the shift of entrepreneurs from one generation to the next. These firms have been successful in establishing a brand in a higher priced segment of the market and by turning direct sales to the traditional one-line retail trade sector. To meet the new requirements of the retail trade, they have established regional order centres in the different urban agglomerations of Germany. They also cater for short term small orders, which create new requirements for flexible production. They have not, however, upgraded to the high fashion segment, where Italian competitors are so strong, but look for larger market niches in high quality footwear. Apart from organising their particular distribution channels, this strategy has also required both cost reduction and product development. Consequently, firms have adopted the usual cost reduction strategy of relocating the production of uppers to Hungary and other countries. However, they have, at the same time, expanded their design departments at home.

Kennel & Schmenger is a case in point. Five years ago, the entrepreneur started completely to alter the strategic position of the firm (created in 1917). One of the main steps was to define four clearly different sets of collections in women's footwear, and radically to improve design and marketing in those segments. While labour intensive upper production was sub-contracted to Hungary, the firm increased and rejuvenated its design staff. Learning, however, is now no longer dependent on internal resources alone, but relies on external resources as well. Annually, four guest designers, from for example, Italy and Austria, are employed for a number of months, despite their very high wages. As the sales manager of the firm, Mr Schwab, put it, "It is the philosophy of the firm to market its own models and to adopt and tune international trends with the philosophy of the firm" (personal interview). He emphasised that the firm's models are copied by competitors and, consequently, he viewed the firm as a leader in its market segments. Additionally, the firm paid one employee living in Italy to collect information on new fashion trends and on suppliers. Another source of competitiveness was the selection of distribution channels, i.e. of specialist one-line retail shops. Finally, the firm was able to make use of modern process technology, such as CAD and CNC, through leasing rather than buying. This increases flexibility in production. "We are able to start on the client's request quickly, to translate new trends immediately and to deliver without delay" (interview with Mr Kennel and Schulhoff, 1997).

These firms are very interested in improving the knowledge base in the region (vocational school, training, college) and make extensive use of the public research laboratory, the remaining local suppliers and machinery industries. To give an example of the latter relationships, some local footwear producers can improve their machines with the assistance of local mechanical engineering firms while these firms can demonstrate their new products to clients 'at work' in a nearby footwear factory. More important seem to be public institutions in knowledge production and transfer as the small and medium sized firms in the footwear industry are mostly unable to invest in their own departments for development and training. With a strategy of raising product quality and bringing in new process technologies, demand for qualified personnel is rising. Product design, which is absolutely necessary for the firm's capacity to establish a brand and make new collections, is, however, mainly based on a more or less traditional qualification system in the region. The number of freelancers in product design is decreasing. Firms are trying to internalise this activity but increasingly face problems in making appointments.

It has not been clear until now who should provide what qualifications. The Land government of Rhineland-Palatinate recently relocated part of a technical college to Pirmasens and established a department of footwear production - the only one in Germany. This, however, is more orientated towards product and process technologies than to design, and the qualification gained by students seems to relate more to the requirements of large firms than to SMEs, and particularly to the requirements of the traders/importers because they need highly qualified personnel to control their foreign plants and sub-contractors. Training initiatives from other institutions are few. Indeed, recently the association of footwear producers was not prepared to increase the attractiveness of vocational training by establishing a fellowship programme.

Thus, 'modernisers' in the footwear sector have an urgent need for the collective characteristics and benefits of an industrial district, no matter whether they are old or recently established. They need the district as an environment for incremental learning (Asheim 1997). Radical innovations break with knowledge traditions but can make selective use of those traditions at the same time. In the Pirmasens area, local entrepreneurship is still able to create 'radical innovators', albeit rather exceptionally.

In the footwear survey, three firms were revealed as such radical innovators. One was Manfred Ehrgott who developed an ecological shoe model. He was born in Pirmasens but was not 'embedded' in the traditional philosophy of how to make shoes because he had worked as a public wage controller for homeworkers in footwear production. However, he became the manger of an ecological shoe producer for one year. To developing his ideas on how to produce a 'real' ecological shoe, he faced extreme difficulties in obtaining capital from the banking sector for firm creation ('Geo'). He finally managed to find a local partner, a member of an old entrepreneurial family in the Pirmasens footwear sector, who had made his money in real estate. Ehrgott contributed to the development of an ecological leather tanning process by the large chemical firm BASF, and managed to obtain ecological parts. Once the problems of product and process development had been solved, the problem of choosing a proper distribution channel for this kind of high priced footwear had to be overcome. The firm finally distributed mainly to specialist ecological and sports shoe mail order firms. Currently, the firm is attempting to establish its own retail shops in selected cities, typically starting with Germany's 'ecological capital', Freiburg, and on the internet.

Why entrepreneurs in the footwear industry have established their new firms in the Pirmasens region appears to be mostly a matter of private motives (area of origin, marriage, and the like). However, radical innovators have changed almost all areas of activities in footwear production and distribution. Innovation occurs in many ways. One radical innovator started especially flexible production, combining general modules into unique pair assembly in the orthopaedic sector. Others invented new ecological products and introduced new forms of distribution to serve a well-off clientele in urban agglomerations or abroad (notably Japan). Changes in product design technology, through computer design for example, gave them independence from the declining traditional local skills base but also made them attractive to young people. It comes as no surprise that production in these footwear firms is growing.

Conclusions

Despite their different technologies and organisational forms the local wickerwork and footwear clusters studied here have experienced similar trajectories of decline in recent decades. Firms can prise themselves free of local lock-ins only through learning-by-using and creating knowledge on new products and new distribution channels. Firms choosing to leave a declining cluster have had to overcome the problems of sunk costs associated with past experience on products, processes and on markets. Clark and Wrigley have argued that these sunk costs are "embodied in organisational structure and manager's careers" (1997, p.341). From this study it appears that this contention holds true also for owner operators in small and medium sized family firms. Furthermore, a firm needs redundant resources if the entrepreneur is looking for radical changes as a strategy for leaving the cluster. Because past experience can be used as an asset and because most firms have insufficient capital or credit, it is not surprising that a general tendency is for firms to attempt to stay in a sector as far as possible and to 'learn' new routines, new products and new markets incrementally (Asheim 1997).

As a result, exit strategies have been rather similar in both the wickerwork and footwear clusters of SMEs, despite the more restricted exit possibilities in the Lichtenfels/Michelau district owing to the simpler technological *regime* in wickerwork compared to footwear production. Small firms in the consumer goods sectors tended to stay in the industries as long as possible reducing to local niche markets that have grown through tourism. Intermediate producers in supporting and related sectors were sometimes successful in using their knowledge and learning to move into new markets, and left the production system. For example, upholstery manufacturers shifted into plastic production. This exit strategy deserves more detailed analysis beyond the scope of this chapter because of the tremendous impact it can have on the prosperity of a region.

This chapter has demonstrated the diversity of entrepreneurial strategies to cope with sectoral decline in a declining local production system. It is quite obvious that some entrepreneurs can successfully make use of the external assets of an old production cluster to create new external assets that favour their firms. Surprisingly, the state has invested only very recently in the improvement of the technological base in both regions, i.e. in wicker furniture production and high quality shoe production. These examples underline the arguments raised in current scholarly debate on learning regions and the stickiness of production clusters (see, for example, Maskell and Malmberg 1997).

The shift from low-tech mass production in sharply declining sectors to specialised and flexible niche production has resulted in increasing fragmentation and the segmentation of local production systems. The emergence of new sub-clusters in the regions studied here may no longer rely on the same social and technological environment. Moreover, the shift in production has not halted the decline of local production. A recent report on the economic situation of the German footwear industry reinforces this view. The relocation of footwear production to low wage countries is still continuing, leading to the conclusion that the future of the German footwear sector will be in product development and the control of the market (Haid and Wessels 1997).

In fact, this process is clearly demonstrated by the exit strategy of many entrepreneurs and firms to become traders and importers rather than to continue to produce locally. In focusing and extending their knowledge and power on distribution channels, these firms have survived or even grown in their traditional markets but have left production. There are, nevertheless, differences between the two consumer goods sectors considered here. In the rather shallow production system and craft-based organisation of the wicker industry, trading firms no longer show any kind of embeddedness in the local production system. They remain in the area through inertia alone. Traders and importers in the footwear production system are different. They still make considerable use of regional know-how in footwear production, of public institutions for testing (imported) products, of the newly emerging public infrastructure for training and qualification, and of independent footwear designers and collection producers. The regional production cluster has been slowly transformed into a decentralised 'centre of competence' at the heart of non-local production chains. Apart from the fact that these processes do not solve the problems of employment on the regional labour market, it is not clear whether these competencies can remain at all 'sticky' when production has become mobile and is relocated to low-wage countries.

The two consumer goods sectors considered in this chapter are orientated towards niche markets, which have limited export potential. What might be a successful strategy for an alert entrepreneur such as Schütz or Kennel will not be possible for all members of the production system. Regional authorities have long since taken the logical step of trying to diversify regional economic structures but with only partial success. Though manufacturing still persists in both regions so too does unemployment.

12 Modernising the Industrial District: Rejuvenation or Managerial Colonisation?

Bengt Johannisson

Introduction

In spite of many ominous prophecies the importance of small enterprises as innovators and job creators in the economic system is not diminishing. Interfirm personal networking, so typical for small firms, is now considered to be the generic way of organising economic activity in turbulent environments. This increasing concern for networking, in particular when localised, is shared by the business community and societal institutions. Policy measures are increasingly focusing on systems of firms, not individual enterprises. However, these induced strategies for business development emanate from images of knowledge, which have their origin in corporate management practices.

The aim of this chapter is critically to investigate conceptually and empirically how measures which are associated with 'professional management' as an ideologically coloured watchword, may, albeit with the best of intentions, undermine an industrial district that, as a collective, demonstrates entrepreneurial qualities. The theoretical contribution of the chapter draws upon entrepreneurship theory as well as complexity and network frameworks. A Swedish industrial district provides the empirical data. This organically constructed socio-economic texture is constantly under pressure from various institutions' efforts to teach the local business community to become 'knowledgeable' in the 'proper', i.e. managerial, way. These ambitions include the creation of an industrial development centre devoted to an escalation of knowledge creation. The Swedish case will discuss strategies and tactics available for use by the local business

283

community to maintain the unique entrepreneurial features of the industrial district. Lessons for practitioners conclude the chapter.

The challenge: sustaining the 'collaborativeness' of firms and localities

A standard phrase used by both researchers and practitioners in the 1990s is 'we live in a knowledge society and do business in a network economy'. This has put the spotlight on learning associated with business exchange and has triggered a greater concern for applying a spatial perspective on economic development. Maskell et al (1998) provide a recent and thorough European introduction to this perspective; similar thoughts were presented by Storper (1995) (see also Chapter 2 and 8 in this volume). This emerging framework suggests that qualified knowledge, in spite of the IT revolution, remains tacit, and that associated learning is an outcome of interactive human processes. Consequently, economic development is favoured by physical and associated cultural and social proximity. Personal face-to-face contacts between businesspersons (Lundvall 1993), make knowledge transfer (a generic kind of knowledge creation (Nonaka & Takeuchi 1995)) feasible. This image obviously enforces the overall potential of the region as an economic actor.

We share this basic view concerning the viability of spatially organised economic activity. We would nevertheless like to propose a different set of arguments for, first, making the positive dynamics of spatial clusters of firms intelligible and, second, identifying the barriers to an induced regional strategy for economic development. Since most localities are dominated by family businesses it is more feasible to relate to small business behaviour in general and entrepreneurial conduct in particular. This stance includes questioning the image of firms as mainly profit seeking and positioning against competitors. If owner-managers are equally existentially and economically motivated, their objectives vary according to their overall life situation and the location of their firm. If firms are more oriented toward what they do *for* or even *with* their customers than toward concern for what they do *against* their competitors, then collaborativeness and not competitiveness becomes the prime source of success. As suggested by Ring and Van de Ven (1994) both efficiency and equity (fair dealing) guide such collaborative linkages. If transactions are not solely associated with costs but also with arenas for a creative dialogue that can generate business opportunities, the networked resource base and not the individual firm's resources becomes a major determinant of firm performance (see Dyer and Singh 1998).

Different conceptual foci may be applied to the same empirical phenomenon - here inter-firm co-operation in a spatial setting. Combining entrepreneurship, complexity, and network frameworks we hope to be able to shed light from a different angle on proposed sources of spatial small firm clustering and commercial success, such as 'the air' (Marshall 1920), 'untraded interdependencies' (Storper 1995), 'sticky' places (Markusen 1996), and 'localised learning' (Maskell et al 1998, Lorenzen 1998). The approach proposed here provides some special explanations for the alleged inability of entrepreneurial ventures and the business community in industrial districts to absorb codified knowledge and integrate it with experientially acquired local insight. It may not be a matter of ability but rather a matter of not wanting - and for good reasons.

We see four interrelated issues as barriers to recognition of the potential of spatial small firm clusters and consequently to viable and legitimate regional strategies for economic development. *First*, there is a weak understanding of entrepreneurial processes that out of ambiguity create and materialise business opportunities by mobilising resources independent of their origin. Instead, the major concern in the literature is about reducing uncertainty by using existing owned resources within an elaborate institutional framework (North 1990). *Second*, there is an exaggerated belief in the possibilities of controlling economic processes by strategic intent instead of continuous improvisation where both ends and means are alertly co-ordinated according to 'unknowable' environmental changes. *Third*, there is a dominant ideology nurtured by large corporations (managerialism) that denies human 'irrationality', a form of conduct necessary to initiate and complete innovative and entrepreneurial processes. *Fourth*, and in consequence of the first three issues, there is a lack of understanding within the political system of the unique features of spatial entrepreneurship, a lack that increases the risk of misdirected regional public policy. These issues will be dealt with below.

In the next section arguments for recognising entrepreneurship as a genuinely interactive and collective phenomenon are presented and elaborated with respect to the spatial dimension of entrepreneurial processes. In the third section of the chapter the dynamics of such localised entrepreneurship are conceptualised drawing upon complexity theory and demonstrated empirically with network data from an industrial district in Sweden. Most often the term network/networking is used metaphorically, as it also is in the literature on the industrial district as pointed out by Maskell et al (1998). It is therefore important to provide appropriate data, especially since this can help further theorising. In the fourth section, three different ideologies are presented which co-exist within industrial districts dominated by small family-operated businesses. In the following section the learning and organising features of an industrial district are conceptualised in order to provide a platform for a critical review of the implications of an induced endeavour to enhance the capabilities of local firms. In the sixth section this case is presented and interpreted as a contest between different ideologies. The concluding section summarises the lessons gained for academic research, public policy, and the business community.

Entrepreneurship as a collective phenomenon

Our concern here is small business and entrepreneurship in a spatial, regional setting. First a few words about the adopted definitions of these concepts. Most *small businesses* are microfirms with no more than nine employees and are owned and operated by primary family members (see Brunåker 1996). There is no corresponding straightforward definition of *entrepreneurship*. Economists associate the phenomenon with *what* its innovative capacity achieves in the market place, psychologists with *who* the major agent is. The anthropologists and management theorists are concerned with the *whys*, the motives for launching change processes in the market place or in corporate settings. Recent research, though, is more interested in *how* entrepreneurial processes are initiated and how they sediment into new ventures (Stevenson and Jarillo 1990, Gartner et al 1992).

We perceive entrepreneurship to be a matter of *processes whereby resources, independent of origin, are organised into new patterns according to opportunity.* The key to insight into the entrepreneurial phenomenon is then to focus on how this new organising is brought about. This means that entrepreneurial agents are assumed to not only accept risk and cope with uncertainty, but also to even thrive on ambiguity. Entrepreneurship means enacting new realities, more concretely business ventures; the boundaries between the individual initiator of the process, the venture, and the context remain fuzzy. This implies constant changes in existing ventures, including the termination of some, as well as the creation of new ventures. These processes can be made intelligible by way of a network metaphor (Johannisson 1992, Taylor 1999). Entrepreneurship is thus about emergence and constant learning. The entrepreneur's ability to envision, her/his alertness to environmental change, whether structural or coincidental (Bouchikhi 1993), and her/his ability to initiate, improvise, and inspire others, are factors which all influence the outcome of entrepreneurial processes.

Economic and psychological approaches associate entrepreneurship with individual action - with the lonely hero. However, just as with innovation and learning, entrepreneurship is the outcome of interaction (Van de Ven 1993). Here entrepreneurship is presented as a generically social, a *collective,* phenomenon. The conceptualisation of a new venture, making sense of an ambiguous environment, calls for multiple interpersonal dialogues, for 'polylogue'. Such personal networking is congruent with a presumed strong need for independence among entrepreneurs. When network ties are created, parties volunteer to interact. However, due to the mutuality of such ties, once they are established, they create dependence. Since this dependence (and network) is, as much as the venture itself, created by the entrepreneur himself or herself, it is not perceived as a straitjacket.

The subsequent organising of resources within the entrepreneurial venture is based on social and business exchange that has its origin in mutual trust and commitment. The ultimate success is recognition in society, i.e. assimilation of the business into established structures. The new venture thus materialises as part of the personal network of the founder and usually also becomes legally demarcated. Once a business has been launched the need to defend achieved independence means that most owner-managers are resistant to influences from both employees and outsiders. This builds a static patriarchal structure and fosters a centripetal worldview that implies a strong but myopic action orientation. The individual small firm is therefore, on the one hand, a very efficient 'action cell' and, on the other hand, very vulnerable to structural changes. Continued entrepreneurship calls for external networking to balance the inward orientation with an outward, collectivist view. As an innovative organiser the entrepreneur has to be socially skilled.

As a collective, small firms represent a major force. This is proven already by the fact that in most national economies the small-firm sector is a major contributor to employment growth. There is a number of more structured collective forms of entrepreneurship. These originate in the 'personal community' created by the network of the initiator. Family involvement in the typical small firm makes it a generic social unit albeit maintaining an internal orientation. Partnerships and co-operatives often originate in concrete needs for collective action. Forms of deliberately designed collective entrepreneurship include, for example, franchising systems and strategic alliances.

A spatial view comes to the fore when focusing on collective entrepreneurship where the boundaries dissolve between, on one hand, individual ambition and economic action as reflected in the firm's operations and, on the other hand, the social context. Considering the general social embeddedness of economic activity (Granovetter 1985), it is easy to imagine that in a local context instrumental and existential images of entrepreneuring may merge. The Marshallian 'industrial district' then appears as to be a qualified case of collective entrepreneurship. In the industrial district interaction between firms breeds venturing processes spontaneously enacting new firms. In the industrial district the business community as a whole counts; the district thrives on high local volatility in terms of business start-ups and closedowns. The 'irrationalities' of these entrepreneurial processes mean that their features are difficult to grasp using a decontextualised linear logic within which planning and evaluation make sense. Since there is no 'visible hand' guiding these processes a different approach is needed if they are to be intelligible. Complexity and chaos theory or 'chaotics' (Tsoukas 1998) provides such an approach, accommodating both order and disorder.

The industrial district as a self-organising system

Chaotics - an emerging framework

The concept of self-organisation is inherent in the emerging theories of complexity and non-linear dynamics. *Self-organising* develops when autonomous units, capable of multiple 'behavioural choices', operate with individual freedom but within clearly defined frames of reference. The result is coherent sets or patterns, without hierarchy and without implied permanence. Briggs and Peat (1989) describe interactions of chemicals, termites, slime-mould amoebae, human traffic, and the growth of cities as examples of self-organising. Stacey (1993, 1996) and Morgan (1993) provide illustrations from the world of human organising. Weick's (1976) image of 'loose coupling' and MacPherson et al's (1992) elaboration on the need for both strong and weak ties between units add to our understanding of self-organising. In a 'loosely coupled' system the units preserve their identity but integrate as well. This means that the units are mutually responsive as all interacting human beings are.

Emerging self-organised patterns of interaction have *bounded instability* i.e. represent an unstable system within limits. This state of bounded instability implies the coexistence of competition and co-operation, that information both flows freely and is retained. (Stacey 1996, pp. 97-98). In self-organising systems both internal and external boundaries are fuzzy.

The complexity and dynamics of a system, as well as its variability and viability, increase with the number of members. However, the energy demand to maintain any desired configuration also increases. Such self-organised systems have been called *'dissipative structures'* which are "... capable of maintaining their identity only by remaining continually open to the flux and flow of their environment." (Briggs and Peat 1989, p. 139) Continuous change, or constant emergence, is a 'natural state' in dissipative structures as well as in entrepreneurship (Gartner et al 1992) - the change and learning is needed to maintain the identity of the system. While energy is needed to change other structures, dissipative structures need energy in order *not* to dissolve (Stacey 1996). In human systems the energy needed to maintain the system comes from different sources: personal involvement of members, committed agents of members' personal networks, and legitimacy and, less so, resources that can be traded.

Complex or *strange* attractors represent the generic principle, a basic order, which guides the evolution of the complex system. Every viable enterprise has its own complex attractor, which represents *a set of inherent principles guiding a process of evolution with patterned but indeterminate results.* The concrete outcome of change processes is sensitively dependent on initial conditions - that is, regularity and irregularity coexist (see Frank and Lueger 1997). In the individual entrepreneurial firm the founder's vision is guided by, and not only gives guidance to, everyday activities. A forceful vision, an image that makes the firm tick, is intuitively comprehended by both the entrepreneur and her/his associates. Notions such as 'social embeddedness', 'untraded interdependencies' and 'stickiness' suggest an implicit order guiding firms' interaction in spatial small firm clusters. This order represents the complex attractor that defines the limits within which the system must be kept to maintain its viability.

Industrial districts consist of internally well integrated, yet independent, businesses which collectively demonstrate great potential for self-organising. There are three main reasons for this. First, the interaction between firms is both calculative and social, which implies variety and subtleness, alertness and concern with respect to business exchange, as well as mutual control. Second, no formal authority exists to impose standardised action - even if informal institutions provide a set of rules

(North 1990). Structure is the outcome of spontaneous processes and the social embeddedness of the economic activities acquaints enculturated district members with the local rules of the game, the 'complex attractor'. Just as the recognition of emerging opportunities needs a prepared mind, self-organising is the outcome of general responsiveness within a strong frame of reference (Wheatley 1992). Third, a strong collective-self-confidence provides the courage to keep external boundaries fuzzy, thereby maintaining the dissipative structure of the industrial district (see Dandridge and Johannisson 1996).

Operationalising interactive complex systems - the networked industrial district

Even if used solely metaphorically, the emerging chaotics framework offers a way of dealing with the dynamics and complexity, ambiguity and unpredictability. To bridge this metaphor and economic life as practised, for example in an industrial district, calls for more substantive modelling. Business systems as human interactive systems with their nodes and linkages offer a huge potential complexity if not confined by the straitjacket of an intended structure. The ideal (i.e. supposedly 'efficient') *hierarchy*, given a span of control of 8, only needs 72 relationships to organise 73 persons. Between them, these 73 people may, however, establish 2,628 (73 x (73-1)/2) undirected linkages emanating from any instrumental need, shared interest or mutual feelings (Sjöstrand 1992). If this potential is taken full advantage of, information fed into any node of such a system will soon disperse throughout the system at large. This has two important implications. First, the socio-economic *networked structure* represents a much larger potential for innovation than the hierarchy. Not only are there vertical ties, as in the hierarchy, and horizontal ties, often a basis for alliances between competitors, but also there are 'lateral' linkages which deny any preconceived categorisation of nodes and instead enact new combinations (Burns and Stalker 1961). Second, no member of a network with more than, say, 25 members can consciously keep all optional ties to and between others in mind. Consequently, members have to rely on intuition and tacit knowing, making the system at large self-organising.

In proposed self-organising spatial small business systems, such as the industrial district, networking generally becomes personalised and comprehensive. Exchange between firms is then both the origin and the outcome of commitments between individual persons who represent themselves as much as their firms. The viability of firms is dependent upon the individual owner-manager's capability and credibility as reflected in her/his personal network.

Our research in Sweden throws light on the nature of these personal networks in different locational contexts. In this research into small-firm clusters and collective entrepreneurship, two spatial business systems have been studied: a community within an industrial district and a science park at a university. In these business settings the sociocentric networks integrating owner-managers and their firms are identified. The data presented below are more extensively reported in Johannisson et al (1994). The sample responding to a mail survey includes 67 firms from the community and 42 firms from the science park. Since the social and business dimension of personal networks were each specified as twofold - *talk/acquaintance* and *commercial/professional* respectively - four potential strands of relationships to other local firms were identified as follows:

The empirical analysis shows that while only two of the firms in the industrial district had no commercial ties to peer establishments, half of the firms in the science park were such 'isolates'. By studying the density of the network composing the commercial network we can go beyond the node level in analysing the networks. Our data then showed that while in the industrial district 14.4 per cent of all potential ties were realised - i.e. 318 out of 2,211 (67x66/2) - only 2.6 per cent were used in the science park. When analysing the acquaintance network, the difference was even more dramatic: 31.3 per cent realised for the industrial district versus 3.5 per cent for the science park. Obviously self-organising processes, generated by and reproducing these network structures, were much more likely in the industrial district than in the science park.

SOCIAL

Talk   A face-to-face or telephone meeting was held with the senior management of the firm over the last 30 days. The conversation should have lasted for at least five minutes and concerned things other than the weather.

Acquaintance   The CEO or anyone else in senior management and on the board is personally known.

BUSINESS

Some business has been transacted (including lending, borrowing, and barter) with the firm over the last nine months.

The staff of the firm is approached if an ingenious or challenging problem turns up.

The analysis of dyadic interaction does not, though, provide a more comprehensive or holistic image of variations in business system complexity. Computerised graph analytical toolkits make more advanced network modelling possible, presenting e.g. structural concepts such as 'blocks' and 'cliques'. A *'block'* is a network substructure where there is no 'cutpoint', i.e. a single node on which the completeness of the network depends, as is the case where there is a single 'broker' or 'bridge'. Analysing blocks is an efficient way to consider the fragility/robustness of established networks. Consider for example the commercial network in the industrial district, where the block datum says that 63 out of 67 firms are interlinked in such a way that even if one firm closes down, the remaining 62 in the comprehensive network would still be interconnected by business transactions. In the science park, in contrast, the identified robust block structures do not exceed 10 firms.

The above network analysis concerns patterns based on interrelated network nodes - owner-managers and their firms - featuring different images of *'connectedness'* in the network. This represents a potential for collective initiatives where, according to the principles of 'loose coupling', every single tie does not have to be continuously exploited. This potential for self-organising in the industrial district is further reflected in the notion of the *'clique'*. The one-clique is a substructure where all members are *directly* connected. Considering only the largest one-cliques in the acquaintance network in the district there are six, each with fourteen members (Johannisson et al 1994, Table 8). Considering that only 67 members make up the network this clearly proves that the industrial district consists of a number of overlapping substructures. This network redundancy adds significantly to the organising potential that coping with 'unknowable' environment call for (Stacey 1996). Industrial districts, as in this Swedish case, present themselves not only as 'communities of practice' but as 'communities of communities of practice', indeed a sensitive texture for learning and innovation (Brown and Duguid 1991).

The idiosyncratic, complex attractor of a self-organising business system thus does not only reflect direct or indirect (dyadic) ties between specified establishments/leaders but relates to more holistic features of the business system. Further network analysis reported in Johannisson et al (1994) suggests that sub-clusters of firms, defined by similarity as regards the members' linkages to all other district members (i.e. 'structural equivalence'), adopt similar principles in their organising. Such principles deduced from empirical research include a 'balanced' (i.e. neither high, nor low) diversity of members with respect to industry, 'balanced' separation/integration of these members, and a 'balanced' set of internal/external linkages. In the science park, in contrast, most substructures were unbalanced in these respects.

In spite of technical advances, using graph analysis to explore 'self-organisation' in the entrepreneurial networks of industrial districts only provides indications of the subtle socio-economic processes that underpin the viability of the industrial district. Patterns of self-organising, 'complex attractors', define every locality as a unique setting. Policy interventions which do not pay due respect to this may ruin the core of self-organising in a place. In the next section a general threat to such unique localised textures is introduced: management and its claim to provide ubiquitous knowledge and *the* way of ordering economic activity.

Managerialism as an invasive ideology in the industrial district

Entrepreneurship is about the creation and enactment of new ways of structuring resources and organising stakeholders. Shared features of entrepreneurship and industrial district self-organisation include for example playfulness/membership in local associations, improvisation/ flexibility in local business exchange, commitment/local identity, and social capabilities/community membership. These features are not only disregarded in corporate settings, they even challenge corporations and their instrumental and hierarchical order. This difference in kind between entrepreneurial and corporate conduct is critical because it is not just a matter of different (cognitive, analytical) perspectives on reality, as is the case with the notion of (business) logic (see, for example, Prahalad and Bettis 1986). It is rather a matter of contrasting *ideologies.*

Ideology is here defined *as a taken-for-granted way of perceiving, evaluating, and acting upon reality thereby mobilising cognitive as well as emotional forces.* An ideology becomes self-enforced by way of focused perceptions and strong commitments. It is only natural for its proponents to impose their own image of reality on others.

Managerialism as the ideology originating in the corporate world has, according to Deetz (1992), colonised not only the business community but also society at large. Elsewhere we have presented *entrepreneurialism* as a counter-ideology to management (Hjorth and Johannisson 1997). As the argument goes, entrepreneurship cannot be made intelligible within a management perspective (Bygrave 1989). Managerialism and entrepreneurialism as sense-making and action-triggering devices represent two phenomena different in kind.

Table 12.1   The family business as the scene of action for contrasting ideologies

	Entrepreneurialism	Paternalism	Managerialism
Structuring of activities	Organic (continuous boundaryspanning networking)	Clan (emotional hierarchy)	Mechanistic (functional hierarchy)
Resource control	Trust relationships	Proprietary ownership	Contractual ties
Image of the market	Ambiguous, potential collaborative space	Familiar, screened-off space	Uncertain, risky competitive space
Time perspective	Emergence	Past, present, and future	Present and future
Core competencies	Tacit holistic and associative knowledge	Tacit holistic and local knowledge	Formal linear and instrumental knowledge
Success criterion	Resources for maintaining venturing processes	Ensuring the family empire	Return on investment and quantitative growth

As stated earlier the industrial district as an empirical representation of collective entrepreneurship emerges out of owner-managed family businesses. Therefore, we have to introduce yet another ideology - the family as a social institution. As a dominantly emotional structure the family obviously brings a new perspective into 'business making', whether compared with the stewardship of management or with the new-venturing/innovativeness of entrepreneurship. According to Berg (1979, pp. 256-7) the emotional structure of an organisation orders emotional processes into a historically determined and stable pattern representing the organisation as a whole.

In Table 12.1 some organisational features of the three proposed ideologies are contrasted. *Entrepreneurialism* means the organic organising of internal and external resources. Networking based on trust relationships means that transaction costs are minimised, learning is stimulated, and business opportunities are realised. The business context is continuously enacted implying that favourable events are amplified by own intentional and synchronised action. An experimental attitude implies short-term projects and frequent follow-ups, with vision as a stretchable frame of reference. The emergent character of the venture calls for timing of external events and its own actions. Actions are intuitively directed and integrated by way of analogy and metaphor which to outsiders appears as constant improvisation. Accordingly, successful entrepreneurialism means mobilising enough internal and external resources, including financial capital, to keep the venturing career going.

Paternalism as it appears in a business setting presents itself as a clan structure where the hierarchy is structured by seniority and ties of blood. In this emotionally embedded structure legal ownership of resources, not the services they accommodate (see Penrose 1959/95), is crucial: it reflects the basic property rights upon which all Western societies are founded. The meaning of business life is the creation of a safe domicile for the family (Miller and Rice 1967). This implies claiming a controllable niche for the operations of the firm, mentally restricted by tradition and often spatially demarcated as well. In this context, everyday life becomes as important as maintaining traditions and building a future for generations to come. The competencies needed for this endeavour are deeply embedded in the personal histories of the family members and of further confidants inside the firm. Keeping the business within the family is the dominant objective.

In the *industrial district* paternalism and entrepreneurialism coexist symbiotically. This means that at the firm level the paternalistic ideology dominates while in the industrial district as a whole entrepreneurialism is reflected in the way individual actors interact. We have already introduced personal networking as a generic way of depicting features usually ascribed to an industrial district as a viable holistic socio-economic unit.

Managerialism proposes that activities should be structured according to functional areas and that planning and execution are separated, usually both in time and space. Linear-analytical decision rationality rules. Internal and external dependencies are controlled by contractual agreements. Uncertainties in the environment are further reduced by way of various institutional arrangements such as adherence to industry and professional norms and the adoption of a management vocabulary. External points of reference (e.g. due to public ownership) mean that images of the present and promises concerning future success have to be continuously communicated. Imagining that the environment is knowable, albeit evasive, means that superior (information) technology is assumed to provide needed competitiveness. Managers continuously have to prove themselves as reliable agents by providing owners with returns on their investments (ROI) that, in combination with quantitative, internal growth, is the indicator of success.

The very notion of 'ideology' signals that entrepreneurialism, paternalism, and managerialism embrace more than individual actors, individual firms, and individual localities. Ideologies coexist in society at large and, therefore, every family business setting has to host all three ideologies in order to endure. Increasingly, institutionalised and internationalised business environments demand managerial reasoning and values. Family values persist in most cultures and increasingly competitive business environments call for entrepreneurial vision and continuous learning.

In contrast to, for example, Hoy and Verser (1994) and Watson (1994), we do not think that ideological tensions can be dealt with by way of some kind of puzzle solving or integrative device. Such a functional 'strategy' for coping with coexisting ideologies would mean submitting to the still dominant managerial approach. Instead, the very tensions between on one hand entrepreneurialism/paternalism, and on the other hand, managerialism have to be integrated in a framework which aims at making the viability of entrepreneurship, family business and localised small firm clusters intelligible. The challenge is to find ways to capitalise upon the energy that these tensions create.

Individual elements of the three ideologies may coincide, though. Paternalism, as a reflection of the family business perspective not only legitimises nepotism, it is also the very reason for founding a business in the first place: it thus relates to entrepreneurialism. Schumpeter's image of entrepreneurship included the vision of founding an empire for generations to come - besides the joy of creating and the need to struggle and succeed (Schumpeter 1911/1934, p. 93). The survival of the firm is recognised as a goal shared by owner-managers and hired management. Penrose (1959/95) points out that entrepreneurship and management take turns in the firm growth process and Czarniawska-Joerges and Wolff (1991) underline that entrepreneurship and management share a concern for organising. These and possibly other shared attributes of the three ideologies do not, however, change the fact that, as holistic sense-making and action-triggering constructs, they differ in kind.

The industrial district as a learning and organising context

The three ideologies introduced above provide different images of adaptation and innovation for learning. According to managerialism there is one generic way of understanding reality and thus only one kind of knowledge to teach those not yet informed. This assumed generalisability and commodification of knowledge makes it feasible to separate training from the everyday practice of doing business and to have experts preach the new message. A specially nurtured managerial vocabulary is as a rhetoric playground continuously changing. Entrepreneurialism instead proposes that the world itself is constantly changing and that learning about it must take place in real time, must become integrated with work, and must adopt everyday language. In the traditional owner-manager context, both the action frame and the everyday language remain as learning and time go by. In order to break that lock-in, internal and external learning must combine (Ylinenpää 1999).

Table 12.2   Managerial and entrepreneurial learning modes

Managerial learning	*Entrepreneurial learning*
Implementational design based on monologue and diffusion of knowledge	Experiential approach originating in dialogue and knowledge creation
Separation of training from work	Integration of work and learning
Professional instruction - formal learning	Peer guidance - social learning
Special international management vocabulary	Everyday spoken language enforced with local technical vocabulary
Circulation and accumulation of ubiquitous and commodified knowledge	Perpetuation of idiosyncratic and localised knowing

Entrepreneurial learning, which embraces both individual firms and small-firm clusters such as industrial districts, adds new dimensions to the traditional small-firm learning. This is shown in Table 12.2 where managerial and entrepreneurial learning are contrasted. Entrepreneurial learning originates in the dialogue, which accompanies the continuous organisation of resources in general, and of human resources in particular.

Enacting new realities means continuous real life experimentation. Empathy and trust are basic elements in the sharing of experiences within both the relationships between fellow entrepreneurs and mentoring relationships between experienced and prospective entrepreneurs. Learning is inseparably related to identity-enforcement, and their use everyday ensures that experience as a holistic construct remains.

The *unique features of an industrial district as an entrepreneurial and learning texture* can now be summarised. Owner-managed firms, whether traditional family businesses or entrepreneurial organisations, are recognised for their founders' need for independence. The very foundation of an independent business is a way of amplifying the owner-manager's personality/identity into a social construct that encompasses other persons. In this way, businesspersons construct their own reality and enact the very environment that conditions the success of their firms (Smircich and Stubbart 1985). As much as the independent businessperson creates her/his own venture by keeping the boundaries to otherness open, s/he is heavily dependent upon already co-opted elements in the socio-economic environment, such as family, friends, and trusted business partners. The personal network of the businessperson appears as a general tool in the reality-construction process. Basically it is impossible to separate the entrepreneur as the representative of her/his venture and the social processes which form the venture into a being.

While the network as a personal community is by definition without boundaries in time and space, as the setting for everyday interaction it is in practice concentrated in time and space (Giddens 1984, Johannisson et al 1994, Johannisson 1996). When businesspeople enact their own construction/business processes they are especially dependent upon the local setting. The dependence is however mutual: the businessperson's local sense-making and associated action adds to the potential of the context for her-/himself and others. In more generic terms, the industrial district with its self-organising features is addressed as an *organising context* from the point of view of the individual businessperson and his or her firm. The following features are ascribed to the organising context.

- The organising context, representing that part of reality which to its members appears to be meaningful and intelligible, is jointly produced by entrepreneurs and their firms as well as other local stakeholders (first-order enactment) (Weick 1979, Daft and Weick 1984).

- The organising context is an instrument for making sense of a not-yet coded local and global phenomena and for taking action accordingly, using the organising context alternately as an amplifier and shock-absorber coping with environmental change (second-order enactment) (Johannisson 1988).
- The organising context is realised through personal networking where relational processes provide external economies of overview that supplement the internal economies of overview of the small firms comprising the business community - where each member of the firm/organising context perceives it as a whole and any member represents the firm/context (see Ouchi 1980).
- In the organising context, business behavioural norms are embedded in the historically generated and taken-for-granted social texture, making it difficult to explicate and thus to copy its unique operating features or complex attractors.
- The organising context as a loosely coupled system of semi-autonomous firms is self-organising through timing of single-loop and double-loop learning, making the context as a whole (i.e. not necessarily the individual firm's) entrepreneurial.

While the self-constituted community of practices in the industrial district evolves through continuous piecemeal innovative activity, radical demand, technology, or institutional changes usually originate outside the region. The majority of economic and technological changes is 'part of the game' and can usually be controlled by the industrial district as a self-organising system. There is however always a risk that the industrial district's self-organising and learning abilities - its entrepreneurial capabilities - erode. Usually this is not the outcome of collective complacency and introversion but a matter of radical changes beyond the scope of the 'complex attractor'. Then the local order creates lock-ins (Grabher 1993b). Paradoxically, it may thus be argued that only if global awareness and external exchange is elaborated will the organising context maintain its potential as an arena for local exchange. Qualified customers are the major source of innovation and the customer profile may make the difference between success and failure. Most of the firms in the Swedish industrial district studied here operate on industrial markets. Schamp (see Chapter 11) and Lorenzen (1998) report on the survival challenges of low-tech consumer-goods small firms and the European furniture industry where the challenge is to find customer demand.

Restricting himself to a functionalist approach, Van de Ven (1993) proposes an infrastructure of entrepreneurship which, in the same way as the organising context, is co-created and influenced by entrepreneurs.

However, 'under-socialisation' (Granovetter 1985), and also the 'under-ideologisation' of Van de Ven's approach, leaves out important sources of tacit knowing. The redistribution of power between different ways of interpreting and acting upon the evolution of markets, addressed in this chapter as different ideologies, represent changes with which infrastructures cannot cope. We have already proposed that managerialism has become increasingly dominant, forcing its view upon organising and learning on industrial districts, for example. In the next section we report on such an apparent managerialistic intervention in a Swedish industrial district.

Sleeping with the enemy: coping with managerial knowledge

Industrial districts in transition - a Swedish case

The industrial district researched here has traditions, which go back several hundred years. The region is recognised as the most advanced industrial district in Sweden and has been labelled by Davidsson (1995), for example, a 'small firm Mecca'. Elsewhere, we have mapped the organising features of a part of this industrial district by way of personal networking between owner-managers (see Johannisson 1987, Johannisson et al 1994, and the earlier sections of this chapter).

In 1999, the industrial district had a population of about 80,000 and a firm population of well 1,500 dominated by small firms in the low-tech plastics and engineering industries. Over the period 1987-1995 its core was the only Swedish local labour market that did not reduce its employment. In spite of its viability, this industrial district has repeatedly been condemned by external observers because of its lack of high-tech industries and academically educated staff. The area had been regarded as just another agglomeration of small firms, of which most were unwilling to grow and formalise their knowledge base. Its business community had resisted the accusations and instead launched its own collective initiatives. In the early 1980s, for example, challenged by the need for the adoption of CAD/CAM technology in the production processes, a foundation aimed at enhancing local competencies was created on the initiative of the owner-managers. The municipal authorities were invited to participate as well and, with the few companies already acquainted with the technology, successfully ran training programmes for their peer owner-managers and their employees. In spite of this and further local initiatives, the accusations of inferior local capacities for business development escalated in the 1990s. Reasons for this included increased focus on formal codified knowledge in business settings in general, and institutional changes in Sweden in particular.

Still, in the mid 1990s the Swedish system for industrial knowledge creation and dissemination was fragmented. The universities, albeit regionally dispersed over the country, were detached from the business community. Practically oriented research was carried out mainly at national institutes organised by industry while public industrial policy was implemented regionally. However, over the last few years of this decade a number of changes have taken place. Universities have become increasingly involved in concrete business development: it has even been legislated that the Swedish universities, in addition to traditional research and teaching obligations, shall actively present themselves as a resource to the regional community in general, and to SMEs in particular.

In 1997, the Swedish government decided financially to support a number (originally 9, and in 1999 11) of regional industrial development centres (IDC) for a period of three years. Each IDC specialises in technologies, which reflect the region's industrial structure, and the IDCs are expected to network between themselves collectively to make up a national asset. Their primary tasks include being brokers between local firms and different external providers of competencies, enhancing technology, and intensifying product development. The IDCs are also expected generally to enhance the growth potential of the regional small firm sector. In practice the IDCs operate as intermediaries between the regional (small) business community, the nearby universities, as well as national and international institutions.

At the time that the state offered subsidies, some of IDCs had existed for a number of years while others were created to take advantage of this financial opportunity. In other locations, the offering of state subsidies was synchronised with local initiatives. The industrial district IDC studied here is of this latter kind, originating among a group of small business owner-managers who wanted to further develop the local CAD/CAM initiative attempted some 15 years earlier. The industrial district IDC was thus constituted in late 1996 and started its operations in the beginning of 1997. It specialises in plastics and cutting technology. The shareholders of the IDC are local businesses - (in April 1999, 69 firms) - and three regional metal workers' labour unions. Numbers of local staff have successively increased and included after two-and-a-half years of operation seven people, some of whom are from the region.

The establishment of the IDC in the industrial district is part of a broader local ambition to widen and deepen the knowledge base in the municipality where it is located. The IDC thus far has been accommodated in a brand new building designed to bring together municipal vocational training. Spectacularly, contrasting local standards and prominently located on a hill, this 'knowledge centre' symbolises the challenge for the prevailing business and learning culture in the industrial district which, as stated, means that working, learning, and innovation are integrated in time and space.

One prominent example of local 'projecting' associated with the IDC is the network programme that materialised late 1997. Four firms in the industrial district have enacted an alliance aimed at 'learning through networking'. The general objective of this project is the development of methods for competence enhancement, and it is associated with the ADAPT EU-programme. The four 'core' firms stand out regionally due the their size (between 250 and 600 employees) and absentee ownership. The collaboration between the four companies is managed by a steering committee, including, besides representatives for the four companies, a union representative, staff members at the IDC, and representatives of two regional universities. The steering committee defines the objectives of the overall project and allocates resources to each project. In addition it offers an arena for discussion and problem resolution.

Five network projects have been initiated: methods for knowledge creation, product development, quality, purchase, and forecasting. The network projects are staffed with representatives of the firms and the IDC. The groups meet about every six weeks. What issues are put on the agenda and how they are dealt with vary between the groups. For example, the 'knowledge group' has visited a larger company outside the region to learn about methods. A major private buyer in Sweden has visited the 'purchase group', and the 'forecast project' has co-opted an academic specialist. Over time the work in the project groups has developed differently owing to their different contingencies. Although the participating firms are generally highly committed, participants seem to lack the time needed. Nevertheless, further networks are emerging.

The four-company network reflects an intended broader (vertical) networking ambition. Each of these firms has a number of regional sub-contractors, for obvious reasons mostly small owner-managed firms. The new insights acquired in the dialogue between the larger firms are expected to be systematically disseminated in each firm's local supplier network.

This is assumed to be an efficient strategy for approaching the small firms' need for competence development.

A tentative interpretation of the induced learning strategy

The industrial district IDC is considered by external observers to be relatively successful (Berggren et al 1998). Nevertheless, it seems to have been met with a general wait-and-see attitude among the majority of the small owner-managed firms within the industrial district. Given the framework outlined above this may reflect a collective awareness that the creation of the IDC implies that managerialism is about to establish a stronghold in the district thereby challenging the established self-organising order. The local small business community may cope with this challenge either offensively or defensively.

Possible *offensive* strategies include, *first*, that the entrepreneurs, well aware of the present hegemony of managerialism in society at large, have decided to co-opt managerialism through the creation of the IDC. There it can be controlled and possibly used to legitimate endogenous organic strategies to enhance competencies. *Second*, the IDC can also be seen as a way of enforcing the functions of the industrial district as an organising context, especially its role as shock absorber when dramatic external changes hit the community. Such a 'hedging-in' strategy would turn the IDC into an arena where alien managerial visitors can be taken care of. A major mission of the local trade association has for a long time been to control the exchange between, on the one hand, the business community, on the other hand, the municipality, other authorities, and other 'intruders' (such as researchers!).

Defensive interpretations of the hesitant reaction amongst the majority of the small firms to the creation of the IDC are more varied. They all suggest that the firms attempt to keep the IDC activities at a distance. First, the very approach to learn new competencies through the involvement of external parties is alien to the industrial district and threatens the delicate mechanisms, which both regulate socio-economic exchange and build a learning capability. The majority of the small firms simply disregard these induced changes that put its 'communities of practice' at stake:

"For it is the organisation's [here the industrial district, *this author's comment*] communities, at all levels, who are in contact with the environment and involved in interpretative sense making, congruence finding, and adapting. It is from any site of such interactions that new insights can be co-produced. If an organisational core [here the IDC and allied medium-sized firms, *this author's comment*] overlooks or curtails the enacting in its midst by ignoring or disrupting its communities-of-practice, it threatens its own survival in two ways. It will not only threaten to destroy the very working and learning practices by which it, knowingly or unknowingly, survives. It will also cut itself off from a major source of potential innovation that inevitably arises in the course of that working and learning." (Brown and Duguid 1991, p. 53)

Second, the partial decoupling of knowledge creation and the practice of doing business makes the IDC activities especially appropriate in the larger local firms where managerialism already rules. There the induced networking is accepted since these firms are usually involved in strategic alliances outside the district. The small firm majority may simply argue that what is going on at the IDC is not their business. Also, observing and reflecting upon the lessons learned by these larger firms buys the small business community time.

Third, the large companies have the time and resources needed to invest in the project. Having the large firms be the first to confront the new approach - as reported in the network project above - saves the time and the effort needed to correct mistakes which may be made as a result of inexperience. As a matter of fact this attitude is supported by not yet published research on Swedish R&D activities. This suggests that there is no positive correlation between firms' co-operation with public authorities and competencies centres on one hand, and firms' investments in R&D or benefits of R&D-realised investments on the other. Close co-operation with nearby customers and suppliers, however, enhances both firm R&D activities and their returns. That is, the generic principle of doing business in the industrial community, everyday exchange with business partners, local and non-local is confirmed as an efficient R&D strategy!

Fourth, the IDC itself is just the tip of the iceberg. Its allied universities and national and international institutes, whether involved in concrete local projects or just legitimising the induced network approach, all advocate managerialism. According to many indigenous owner-managers the creation of the IDC signals an organised attack on their paternalistic ideology. Detachment then appears as an appropriate behaviour.

The impact of the creation of the IDC in this case study should not be exaggerated. The local business culture has been questioned for several decades and a number of ways of coping with that questioning have emerged over time. For example, some *individual actors*, who are committed to the district, bridge entrepreneurialism and managerialism. Several entrepreneurs are members of boards of regional and national organisations such as a national advisory board on SME issues, a university, and a major venture capital corporation. One of the leading regional entrepreneurs, a former mayor in one of the district's municipalities, is the informal co-ordinator of fellow Swedish IDC board chairpersons who regularly meet to exchange experiences. This possibly infuses self-confidence in the region and also enables this industrial district to influence its institutional framework. Also, as mentioned, the IDC is to a great extent locally staffed which means that, whatever the rhetoric of the IDC as an institution may be, the action repertoire of its members pays due respect to the local business culture.

Taming the Trojan horse - lessons for researchers and policy makers

The establishment of an IDC as a strategy for induced knowledge enhancement in the industrial district need to be critically reflected upon by researchers. The optimism demonstrated by, for example, Nonaka and Takeuchi (1995) with respect to learning synergies between tacit and explicit knowledge within and across organisational levels in a corporate setting in this perspective seems somewhat unreflected, if not naïve. Also images of learning synergies as applied to industrial districts, from Marshall (1920) to Maskell et al (1998) appear to be too optimistic. The framework outlined here suggests that allergic reactions between different kinds of generation, distribution, and use of knowledge arise from different ideological origins, which are equally plausible as mutually enforcing outcomes. Researchers must be especially aware of this since researchers themselves, through training as members of a professional community, have much more in common with management than with entrepreneurship - and with associated images of knowing and learning. Without such a critical mind, researchers may well both feed and deliver many Trojan horses to well-functioning arenas.

The Swedish setting offers a situation different from, for example, the Italian industrial districts where much of the economic activity was created by institutional changes favouring small firm activity (see, for example, Goodman 1989). In as much as the local structure is conditioned by external support, self-organising may never be developed. Further public intervention is, therefore, needed to supplement local competencies with 'real' services, as is proposed by Brusco (1992) (see also Bianchi 1998). However, generally to recommend the vitalisation of a spatial system of firms through the injection of institutional support, as suggested by

Lorenzen (1998), for example, may be devastating to whatever self-organising capabilities are left in a district. It is crucial not only what services are provided but *how* they are provided. It has to be realised that whatever external and explicit knowledge is added to the knowledge base of the industrial district, its tacit learning and organisational features must be respected.

Lessons for politicians are that socio-economic systems like the Swedish industrial district can only be supported by an amplification of its unique features. Only through the owner-managers themselves can their and their locality's learning be promoted. This has to be respected by policy makers. They need to realise that it takes a long time to enhance competencies and especially to design a future involvement, which does not entail submission to managerialism. Politicians also have to be aware of that support of institutions, such as that offered by IDCs, may initiate local political processes. For example, in 1999 the chairman of the local trade association in a neighbouring community within the industrial district accused the management of the IDC of having the ambition, jointly with one of the 'core' large firms mentioned above, to centralise all further regional development efforts in the IDC municipality.

The lesson for *small business owners* themselves, whether located in an industrial district or not, is that they must take charge of their own learning processes. This suggests that owner-managers, instead of individually enrolling in courses offered by public and private organisers, should *build coalitions between themselves*. These alliances should be used for both the exchange of experiences and the design of courses and for inviting appropriate specialists. The creation of the IDC may have originally reflected this ambition, as were earlier local initiatives. But, the owner-managers have to stay as alert concerning these issues as they always have been with regard to their commercial activities.

Another strategy to cope with invasive managerialism may be to *join forces with external influential business parties* - suppliers, and customers, and those who probably do not want radical change to disturb already working relationships encompassing both learning and innovation. These external parties can also be used critically to review whatever change is proposed by the IDC. A further suggestion is to make institutions such as IDCs into *instruments for dealing with challenges created by the local business community itself.* Formal knowledge is not only a source of knowledge but also a way for new generations of educated owner-managers to create their own futures in localised family business. Successors may be inaugurated as businesspersons into the local business culture through practice-oriented projects run by the IDC thereby offering them a trial run before they re-enter the community in general and their family business in particular.

Notes

1. I acknowledge the contribution of Michael B. Teitz as co-author of an earlier piece, on which this chapter is based. I am grateful for the information provided by the officials of Small and Medium Enterprise Agency of Japan and Japan Institute for SME Research. A thanks also goes to Koichi Okada at Meiji University, Japan, who provided valuable insights on Japan's SME policy.

2. It should be noted that definitions of SMEs differ between Japan and the U.S. I adopted the definitions defined by each country in this chapter. Japan's Ministry of International Trade and Industry defines SME on a sectoral basis, while there is no single definition agreed by the US SME Administration. In the US, however, the most commonly used definition of SMEs is firms with fewer than 500 employees. In Japan, the most widely applied definition of SMEs is firms with fewer than 300 employees. This definitional difference should be taken into account when comparing the figures. Since the Japanese definition is narrower than the US definition, the actual contrast between Japan and US figures is larger than these figures represent in the article.

3. The notion of industrial networks has been as much inspiring as it has been confusing, about what the term actually represents. In this chapter, 'industrial networks' and 'inter-firm networks' are used interchangeably. They refer to business relationships between firms, in forms of sub-contracting, strategic alliances, research alliances and other kinds of collaboration, which involve the exchange of goods, services and information.

4. See Rostow, 1952 for his analysis on the "take-off" stage of economic development. For a discussion on the analysis of "backwardness" from a Marxist perspective, see Gerschenkron, 1962.

5. Historical legacy, institutional structure and social networks have not traditionally been of concern for economists in analysing an economy. Views from policy makers were shared by a former chief economist of the Small Business Association (SBA) (interview on November 1st, 1993, Washington, DC).

6. While concerns for SMEs traces back to the Meiji period, active policy making from a social policy perspective started in 1918. However, active policies towards SME were first introduced in the early 1930s. (Toyoda 1941; Whittaker 1997).

7. The policy measures can be classified into four categories: 1) measures for structural upgrading and other corporate revamping measures for SMEs; 2) measures for rectifying disadvantages in business activities; 3) measures specifically targeted at very small enterprises; and 4) measures in finance and taxation (Aoyama and Teitz 1996).

8. The Temporary Law on Business Conversion and Adjustment Measures for Small and Medium Enterprises was enacted in 1986, and the Law on Extraordinary Measures for Business Conversion of Small and Medium Enterprises was enacted in 1988 (MITI 1989a).

9. The term, SME (*chusho kigyo*) generally refers to small manufacturers, predominantly represented by inner city, craft workshops (*machikoba*). Retail SMEs are not generally referred to as *chusho kigyo*, but rather as *kouri ten*, a specific term that means independent retail stores.

10. However, this does not imply that Japan lacks policies for SMEs in the service sector. In fact, Japan's policy covers virtually all sectors. Yet, the core of the policy, and research conducted on those policies, largely deal with the manufacturing sector.

11. Interview by author, with Professor Koichi Okada, Faculty of Business Administration, Meiji University, Tokyo, January 12, 1994.

12. See Arita 1990, Kurose 1997, Whittaker 1997.

13. The Economist, 1994. See author's interviews, Meiji University Faculty of Business Administration, and Japan SME Research Institute, Tokyo, January 10 and 12, 1994.

14. The Restructuring Law (*ristora-ho*) passed in late 1993 represents such efforts.

15. With heightened competition and the pressure of globalisation, reliance on keiretsu for parts and supplies among Japanese firms has been gradually declining in the past few decades. For example, in the electronics industry, reliance on keiretsu relationships for parts and suppliers typically runs at around 20-30 per cent.

16. For instance, it is only in 1999 that Japan's lawmakers passed a law making it illegal to specify gender in job advertisements. The actual enforcement of equal opportunity for hiring, however, is expected to remain problematic. An opinion survey has already shown that two-thirds of the firms are unprepared to go along with this change.

17. Nippon Keizai Shimbun, February 1, 1999.

18. The Economics Department at Rikkyo University, Tokyo, for example, has run a seminar on entrepreneurship and managing business start-ups called "creating a company" since 1998. The seminar is co-sponsored by Japan Development Bank's new business promotion division, Tokyo Small and Medium Enterprise Investment Development, as well as private-sector research institutes such as Nikko Research (Nihon Keizai Shimbun 1998b). Numerous other examples exist.

19. 'Window-side' office workers refer to those who have been socially alienated from the rest of the workforce due to a lack of social and performance records. Nominally they have been retained under the lifetime employment scheme, but have been moved to the window-side, away from the centre of office activity.

20. See Mutaguchi 1998; also various reports from Asahi Shimbun, Nippon Keizai Shimbun 1998.

21. Reported in Nihon Keizai Shimbun (1998c, d).

22. The study is part of the research programme carried out by the Danish Research Unit on Industrial Dynamics (DRUID). Further information is available at: [http://www.business.auc.dk/druid/].

23. In this contribution a sub-contractor is defined as an enterprise producing components, parts and sub-systems (or undertaking functional operations on materials delivered by the customer) installed in products sold by the customer.

24. The term 'global supply chain' is, here, taken from Schary and Skjøtt-Larsen (1995). In other contributions the term 'transnational supply chain' is seen.

25. The study comprised 1278 manufacturing enterprises. 541 were classified as sub-contractors. The questionnaire was constructed in collaboration with the Danish Statistical Bureau (DS) and carried out by DS in 1995. It comprised 18 questions on forward relations to the 4 most important customers and 19 questions on relations to the 4 most important suppliers. Further information on the questionnaire and the empirical findings may be obtained at Danmarks Statistik (1997) and Andersen and Christensen (1998).

26. The theory of comparative advantage in international trade was a main stream theory of that period, providing an explanatory framework for the understanding of international trade patterns.

27. In his so-called eclectic theory Dunning (1988) has proposed an appealing theoretical framework for understanding expanding foreign direct investment (FDI). It is based on the exploitation of Ownership advantages (O), Locational advantages (L) and the advantages of Internalisation (I). It is called the OLI-model.

28. See Helper and Sako (1995) for a note on this paradox in term of automakers.

29. A very good example of such a step-wise auditing procedure can be seen by accessing the Internet address: [www.nec.com]. Look for the page 'Flow to initial transaction'.

30. A buying centre consists of representatives from those functions and departments in the enterprise having an interest in procurement of the specific item. In some cases specific production units, R&D units and, for example, quality control people, may join the centre. The composition of the buying centre alters with types of supplies under consideration.

31. According to a Danish study carried out in 1995 (Andersen and Christensen 1998, pp. 48) almost 40 per cent of Danish manufacturing sub-contractors can be characterised as traditional sub-contractors, while 21 per cent are classified as standard sub-contractors. A little more than 30 per cent are classified as strategic development suppliers. A small percentage could not be classified.

32. Organisational communities are defined as "a set of co-evolving populations linked by ties of commensalism and symbiosis. Commensalism refers to competition and co-operation between similar units, whereas symbiosis refers to mutual interdependence between dissimilar units" (Aldrich 1999, p. 298).

33. The term impannatories was introduced by Beccatini (1987) denoting a supra-structural tier of firms and specialist actors aiming to co-ordinate activities and promote access to markets for joint products. In this contribution impannatories may be large enterprises providing market access though their global sales or production networks. A case in point is IKEA configuring major parts of the Danish furniture industry.

34. It is thus interesting that Powell and Smith-Doerr (1994) treat these two types of networks in separate sections.

35. This example is taken from the Stainless Steel District in Jutland, Denmark. Three TNCs have invested in the area. By way of their daughter units they have configured more than 160 enterprises and more than 5,500 employees in the area. Two levels of sub-contracting exist. The first level is very formalised and involves 3-4 system suppliers capable of configuring sub-contractors at the next level.

36. Acknowledgements: I would like to thank Meric S. Gertler, Ernst Giese, Eike W. Schamp for their support of this research project. I am especially grateful to Clare L. S. Wiseman for carefully editing the manuscript and pointing out missing links in my line of arguments. This chapter has also benefited from helpful comments and suggestions of Mike Taylor and Eirik Vatne. In addition, I would like to thank the *Deutsche Forschungsgemeinschaft* (German Science Foundation) for providing the funds for this research project.

37. This section is based upon Bathelt (1990).

38. According to Maskell and Malmberg (1998), globalisation processes tend to weaken such regional contexts and threaten the competitiveness of the respective firms because they serve to make some localised capabilities (such as state-of-the-art technologies and organisational designs) more ubiquitous; that is, available at the same cost in may word regions. A similar process that undermines the competitiveness of a firm or region is the codification of tacit knowledge. Initially, newly created knowledge is, at least partially, tacit in that it is specific to those people who learn it and those places and environments where the learning processes take place (see also Conti, this volume).

39. See also (Bathelt 1994, 1995a, b, 1999).

40. The chemical industry is a good example for study. The chemical industry is one of the largest sectors in German manufacturing and is strongly oriented towards international markets. The industry, unlike other sectors (e.g. automobiles), is characterised by an extremely heterogeneous product and process structure (Streck 1984, Amecke 1987, Chapman 1991, Muller-Furstenberger 1995, Bathelt 1995a). Such a heterogeneous sector provides a better study basis than a relatively homogeneous industry group because it is more likely to encompass the multiplicity of strategic responses to the Fordist crisis. In consideration of its heterogeneity, three branches have been selected for a broader analysis of the chemical industry; basic chemicals, pharmaceuticals and PDPV (Bathelt, 1997, 1999). They represent a variety of different configurations of size, growth performance and potential, market structure and pattern of competition.

41. The evolution of particular industry agglomerations and the resultant linkage patterns will not be discussed here. I concur with Krugman (1991) that 'history matters' and will regard the existing spatial pattern of the industry to be a consequence of historical development process.

42. It should be noted that firms usually did not give the permission to quote from the interviews and that it was sometimes not even allowed to record them on tape.

43. Those firms, which serve a limited local market, are among the smallest in the PDPV industry. They do not pose a threat to the market position of larger producers and have, thus, been tolerated over a longer time period (Howard 1996). Their problems in overcoming the Fordist crisis may not necessarily be as great as suggested by Conti (this volume).

44. In addition, not all changes towards flexibility within the product and process structure have been a consequence of the Fordist crisis. There is a long tradition of restructuring processes in the German PDPV industry and other chemical branches. Due to changes in international market regulations, the discovery of more efficient production methods and the consequences of World Wars I and II, German producers have been forced to restructure their production programmes, in parts, from the production of mass products towards speciality chemicals (Schall 1959).

45. The importance of conventional over flexible machinery is supported by Clark and Wrigley's (1997) argument about the functional value of capital and competitiveness. They argue that the value of a firm's machinery is not adequately reflected in its market value. Over time, specific knowledge of the production process and the market is incorporated into the machinery. The machinery becomes increasingly site and plant specific in how it is used. Its importance to the firm may increase over time since this specificity cannot be easily reproduced by potential competitors. A firm's business success may thus be based on the efficient use of conventional machinery, as opposed to a high degree of technological sophistication. In such a situation, conventional technologies would not be replaced by flexible ones.

46. The organisational clusters will also be related to Christensen's (this volume) typology of small firms, which is based on the degree of co-ordination and interactive task complexity in subcontracting relationships. This typology differentiates between *standard* subcontractors, *capacity* subcontractors, *expanded* subcontractors, *strategic* subcontractors and *partnership-based* subcontractors.

47. For the same reasons, PDPV producers try to establish long-term relations with a relatively fixed set of suppliers. To switch from one supplier to another would usually result in a change in input materials, which may have a different chemical composition. From the view of a producer, this would serve to increase adjustment costs and reduce problem-solving capacities that have been acquired in the past. Pure market relations which are based on price considerations are, therefore, less frequent on the supply side than longer-term contacts.

48. Trust relations on the supply side are also established as a means to secure long-term competitiveness. Long-term supplier relations reduce the risk of failure when adjusting the production programmes and processes to meet environmental regulations. Conventional, specialised PDPV producers rely particularly on close supply-side linkages and the problem-solving capacities of their suppliers.

49. Two firms reported that their major customers in the automobile industry had forced them to disseminate firm-specific improvements in the delivery process to other PDFPV suppliers to be used as a new industry standard. This way a codification of tacit knowledge took place, which reduced the firms' innovation success.

50. The term cultural proximity (Gertler 1993, 1996) refers to a uniform set of norms, rules, standards, regulations and experiences which shape industrial practices and organisation, the training system, employment relations, the division of labour, labour participation and attitudes towards technological change and other aspects of the capital-labour-nexus. Within this context, the same language and cultural traditions are only some of the factors that account for nearness. Cultural proximity is often related to the notion of the nation-state.

51. Flexible, specialised PDPV firms correspond with the upper-end subcontractors in Christensen's (this volume) typology (i.e. strategic and partnership-based subcontractors). In contrast, conventional, specialised firms with a spatial specialisation strategy have some resemblance with his lower-end groups of *capacity* and *expanded* subcontractors while those PDPV producers with a product-based specialisation strategy fall somewhere in between.

52. Due to the requirements for and difficulties associated with product adjustment, good long-term customer relations can be quite beneficial. Both sides, customers and producers, know exactly which product specifications are to be met. They also know what to expect in terms of problems that can occur in the adjustment process and how those could be solved.

53. From a user perspective, I would agree with Christensen (this volume) that the acquisition of PDPVs is not a regional phenomenon and that international subcontracting relationships have become increasingly important. My survey results indicate, however, that near-by customers can be quite important for PDPV producers in acquiring specialised problem-solving competencies to serve as a basis of their competitiveness.

54. A similar observation has been made by Koschatsky et al (1996) in a general study of supplier and customer linkages in the Frankfurt region (FRG). Some of the suppliers surveyed predicted decreased competitiveness if the industrial basis of the region is subject to further deindustrialisation tendencies.

55. Research financed by European Community's (DG XII) Targeted Socio-Economic Research Programme (TSER) under the project 'Strategies and Policies for Systemic Interaction in Europe' (CONVERGE).

56. Regional clusters denote the clustering of interdependent firms in one or a few adjacent industries in specific geographical locations.

57. Industrial districts are specific kinds of regional clusters. The term is to describe clusters in the Third Italy, though clusters are found elsewhere. We see industrial districts as regional clusters where industrial activity is supported by specific socio-cultural conditions (i.e. territorially embedded Marshallian agglomeration economies).

58. The three case studies form part of a larger research project to analyse innovation and learning in ten regional clusters in Norway, financed by the Norwegian Research Council.

59. In Jæren, the studied firms belong to a network organisation, TESA (Technical Co-operation).

60. Innovative firms include firms who have brought to market new or improved processes or products in the three years 1995-97, according to the Norwegian section of the Community Innovation Survey.

61. The numbers refer to 1990 (Isaksen and Spilling 1996).

62. In 1999 Rolls Royce bought Vickers, and Ulstein's Norwegian activities are now an integrated part of the RR corporation.

63. At the time of the interviews the yard was owned by Kværner. Now it has been bought back by local entrepreneurs and may lose access to competencies inside Kværner.

64. This take-over has resulted in the production of bicycles being closed down (spring 2000) in Jæren and transferred to Sweden.

65. An earlier version of this chapter was publish in French in Revue International PME 9 (3-4), 1996.

66. Finland and Sweden were members of the EU from the beginning of 1995. Norway is integrated into the internal market through the EES agreement.

67. In terms of the NACE the firms belong to sectors 20, 22, 24 and 27-36. In terms of the ISIC the respective sectors are 331-332 and 381-385.

68. In a multivariate setting the main conclusion will still be almost the same.

69. The share of own products in total turnover is more than 50 per cent.

70. The interviews included questions on the quality of local business services, availability of educated manpower and accessibility. Entrepreneurs' views accorded with a priori views on hierarchical differentiation in local milieus and access to resources.

Bibliography

Aglietta, M. (1979), *A Theory of Capitalist Regulation: The US Experience*, (translated by David Fernbach), NBL, London.

Aldrich, H. (1972), 'Technology and organizational structure: a re-examination of the findings of the Aston Group', *Administrative Science Quarterly*, vol. 17, pp. 26-43.

Aldrich, H. (1999), *Organizations Evolving*, London, Sage Publications.

Allen, J. (1997), 'Economies of power and space', in Lee, R. and Wills, J. (eds) *Geographies of Economies*, London Arnold, pp. 59-70.

Alvstam, C.G. (1996), 'Spatial dimensions of alliances and other strategic manouvres.' in Conti, S., Malecki, E. J. and Oinas, P. (eds), *The Industrial Enterprise and Its Environment: Spatial Perspectives*, London, Avebury, pp. 43-55.

Amecke, H.B. (1987), *Chemiewirtschaft im Überblick. Produkte, Märkte, Strukturen*, VCH, Weinheim.

Amin, A. (1989), 'Flexible specialization and small firms in Italy, myth and realities', *Antipode*, 21(1), pp. 13-34.

Amin, A. (1993), 'The globalization of the economy: an erosion or regional networks', in Grabher, G. (ed.) *The Embedded Firm: On the Socioeconomics of Industrial Networks*. London, Routledge, pp. 278-95.

Amin, A. and Cohendet, P. (1999), 'Learning and adaptation in decentralised business networks', *Environment and Planning D: Society and Space*, 17, pp. 87-104.

317

Amin, A. and Robins K. (1990), 'The re-emergence of regional economies? The mythical geography of flexible accumulation', *Environment and Planning D: Society and Space*, 8(1), pp.7-34.

Amin, A. and Thrift, N. (1997), 'Globalization, socio-economics, territoriality', in Lee, R. and Wills, J. (eds), *Geographies of Economies*, London, Arnold, pp.147-157.

Amin, A. and Thrift, N.J. (1992), 'Neo-Marshallian Nodes in Global Networks', *International Journal of Urban and Regional Research*, 16, pp. 571-587.

Amin, A. and Thrift, N.J. (1994), 'Living in the global', in Amin, A. and Thrift, N. (eds), *Globalization, Institutions and Regional Development in Europe*, Oxford, Oxford University Press, pp. 1-22.

Amin, A. and Thrift, N.J. (1995), 'Territoriality in global political economy', *Nordisk Samhällsgeografisk Tidskrift*, 20, pp. 3-26.

Amsden, A.H. (1989), *Asia's Next Giant: South Korea and late industrialization*, Oxford, Oxford University Press.

Andersen, P.H. and Christensen, P.R. (1998), *Den globale utfordring – Danske underleverandørers internasjonalisering*, København.

Aoyama, Y. (1996), 'Local Economic Development or National Industrial Strategy: Recent Trends in SME Policy in Japan and the United States', *Review of Urban and Regional Development Studies*, vol. 8(1), pp. 1-14.

Aoyama, Y. (1999), 'Policy Interventions for Industrial Network Formation: Contrasting Historical Underpinnings of the SME Policy in Japan and the United States', *SME Economics*, vol. 12(3), pp. 217-231.

Aoyama, Y. and Teitz, M.B. (1996), *Small Business Policy in Japan and the United States. A Comparative Analyses of Objectives and Outcomes*, Institute of International Studies, International Affairs Policy Paper No. 44, Berkeley.

Arita, T. (1990), *Sengo nihon no chusho kigyo seisaku*, Tokyo, Nihon Hyoronsha.

Arita, T. (1997), *Chusho kigyoron: rekishi, riron, seisaku*, Tokyo, Shimpyoron.

Asheim, B.T. (1992), 'Flexible Specialication, Industrial Districts and Small Firms: A Critical Appraisal', in Ernste, H. and Meier, V. (eds), *Regional Development and Contemporary Industrial Response: Extending Flexible Specialisation*, London, Belhaven Press, pp. 34-63.

Asheim, B.T. (1996), 'Industrial Districts as Learning Regions', *European Planning Studies*, vol. 4, pp. 379-400.

Asheim, B.T. (1997), '"Learning regions' in a globalized world economy: towards a new competitive advantage of industrial districts?', in Taylor, M. and Conti, S. (eds), *Interdependent and Uneven Development: Global-Local Perspectives*, Aldershot, Ashgate, pp. 143-176.

Asheim, B.T. (1998), *Learning Regions as Development Coalitions: Partnership as Governance in European Workfare States?* Paper presented at the Second European Urban and Regional Studies Conference on 'Culture, place and space in contemporary Europe', University of Durham, UK, 17-20 September 1998.

Asheim, B.T. and Cooke, P. (1998), 'Localised innovation networks in a global economy: A comparative analysis of endogenous and exogenous regional development approaches', *Comparative Social Research*, vol. 17, pp. 199-240.

Asheim, B.T. and Isaksen, A. (1997), 'Localisation, agglomeration and innovation: towards regional innovation systems in Norway?' *European Planning Studies*, vol. 5(3), pp. 299-330.

Attali, J. (1975), *La parole et l'outil*, Paris , P.U.F.

Attiyeh, R.S. and Wenner, D.L. (1979), 'Critical mass: Key to export profits', *Business Horizons*, December, pp. 28-33.

Audretsch, D. (1995), *Innovation and Industry Evolution*, MIT Press, Cambridge.

Averitt, R.T. (1968), *The Dual Economy*, Morton, New York.

Bathelt, H. (1994), 'Die Beddeutung der Regulationstheorie in der wirtschafts-geographischen Forschung', *Geographische Zeitschrift*, vol. 82, pp. 63-90.

Bathelt, H. (1995a), 'Global competition, international trade and regional concentration: the case of the German chemical industry during the 1980s', *Environment and Planning C: Government and Policy*, vol. 13, pp. 395-424.

Bathelt, H. (1995b), 'Der Einfluß von Flexibilisierungsprozessen auf industrielle Produktionsstrukturen am Beispiel der Chemischen Industrie', *Erdkunde*, vol. 49, pp. 176-196.

Bathelt, H. (1997), *Chemiestandort Deutschland. Technologischer Wandel, Arbeitsteilung und geographische Strukturen in der Chemischen Industrie*, Edition Sigma Rainer Bohn, Berlin.

Bathelt, H. (1999), 'Persistent structures in a turbulent world: the division of labor in the German chemical industry', *Environment and Planning C: Government and Policy*, vol. 17.

Becattini, G. (1978), 'The development of light industry in Tuscany: an interpretatnion', *Economic Notes*, vol. 3, pp. 107-123.

Becattini, G. (1987), *Mercato e forze locali: il distretto industriale*, Bologne, il Mulino.

Becattini, G. (1990), 'The Marshallian industrial districts as a socio-economic notion.' in Pyke, F., Becattini, G. and Sengenberger, W. (eds), *Industrial Districts and Inter-Firm Co-operation in Italy*, International Institute for Labour Studies, Geneva, pp. 37-51.

Becattini, G. and Rullani, E. (1996), 'Local systems and global connections: the role of knowledge', in Cossentino, E., Pyke, E. and Sengenberger, W. (eds), *Local and Regional Response to Global Pressure: The Case of Italy and Its Industrial Districts*, Research Series 103, Geneva, International Institute for Labour Studies, pp. 159-174.

Behlke, J. (1995), *Alte Industriestrukturen und 'Neue Industriedistrikte'. Zur Kritik des Modells am Beispiel der Haustextilienindustrie Deutschlands*, Peter Lang, Frankfurt am Main.

Benecke, F. (1921), *Die Korbflechtindustrie*, Oberfrankens, Leipzig.

Benson, J. (1975), 'The interorganizational network as a political economy.' *Administrative Science Quarterly*, vol. 20, pp.229-248.

Berg, P.O. (1979), *Emotional Structures in Organizations. A Study of the Process of Change in a Swedish Company*, Studentlitteratur, Lund.

Berger, S. and Piore, D. M. (1982), *Dualismo economico e politica nelle società industriali*, Il Mulino, Bologna.

Berggren, C. Brulin, G and Gustafsson, L.G. (1998), *Från Italien till Gnosjö*, Rådet för Arbetslivsforskning, Stockholm.

Bertram, H. (forthcoming), *Kundennahe und Produktionsstandort*, Habilitation thesis, University of Frankfurt/Main.

Bertram, H. and Schamp, E.W. (1989), 'Räumliche Wirkungen neuer Produktionskonzepte in der Automobilindustrie', *Geographische Rundschau*, vol. 41, pp. 284-290.

Bertram, H. and Schamp, E.W (1998), *Deindustralization, Restructuring and Exit Strategies of Firms in Old Industrial Districts: The Case of Footwear and Leather Goods Industries in Germany*. Paper for the 6th Dutch-German Bilateral Seminar on Economic Geography, 7-10 Oct. 1998, West-Terschelling, Netherlands.

Bianchi, G. (1998), 'Requiem of Third Italy? Rise and Fall of a too Successful Concept', *Entrepreneurship and Regional Development*, vol. 10(2), pp. 93-116.

Bijmolt, H.A. and Zwart, P.S. (1994), 'The impact of internal factors on the export success of dutch small and medium-sized firms', *Journal of Small Business Management*, vol. 32(1), pp. 69-83.

Bilkey, W.J. and Tesar, G. (1977), 'The export behavior of smaller-sized Wisconsin firms', *Journal of International Business*, spring/summer, pp. 93-98.

Birch, D.L. (1987), *Job Creation in America: How Our Smallest Companies Put the Most People to Work*, The Free Press, New York.

Birch, D.L. and MacCracken, S. (1983), *The SME share of job creation*, MIT Program on Neighborhood and Regional Change, Cambridge MA.

Birou, L. and Fawcett, S.E. (1993), 'International Purchasing: Benefits, Requirements & Challenges', *International Journal of Purchasing and Materials Management*, Spring, pp. 28-37.

Blau, P.M. (1968), 'Interaction: Social exchange' in Sills, D.I. (ed), *The International Encyclopedia of the Social Sciences*, MacMillan, New York.

Blenker, P. and Christensen, P. R. (1995), 'Interactive strategies in supply chains - a double-edged portfolio approach to small and medium-sized sub-contractors position analyses.' *Entrepreneurship and Regional Development*, vol. 7, pp. 249-264.

Bolton, J.E. (1971), *Small Firms, Report of the Committee of Inquiry on Small Firms*, London, HMSO.

Boswell, J. (1973), *The Rise and Decline of Small Firms*. London, Allen & Unwin.

Bouchikhi, H. (1993), 'A Constructivist Framework for Understanding entrepreneurship Performance', *Organization Studies*, vol. 14(4), pp. 549-570.

Bouwen, R. and Steyaert, C. (1990), 'Construing organizational texture in young entrepreneurial firms', *Journal of Management Studies*, vol. 27(6), pp. 637-649.

Boyer, R. (1990), *The Regulation School: A Critical Introduction*, Columbia University Press, New York.

Braczyk, H.J. and Heidenreich, M. (1998), 'Regional governance structures in a globalized world,' in Braczyk, H.J. et al (eds), *Regional Innovation Systems*, UCL Press, London, pp. 414-440.

Briggs, J. and Peat, F.D. (1989), *The Turbulent Mirror*, Harper and Row, New York.

Brown, J.S. and Duguid, P. (1991), 'Organizational Learning and Communities-of-Practice: toward a Unified View of Working, Learning, and Innovation', *Organization Science*, vol. 2(2), pp. 40-57.

Brunåker, S. (1996), *Introducing Second Generation Family Members into the Family Operated Business - a Constructionist Approach*, Dissertation, Uppsala, Swedish University of Agricultural Sciences.

Brusco, S. (1982), 'The Emilian model: productive decentralisation and social integration', *Cambridge Journal of Economics*, vol. 6, pp. 167-184.

Brusco, S. (1989), 'A policy for industrial districts', in Goodman, E., Bamford, J. and Saynor, P. (eds), *Small Firms and Industrial Districts in Italy*, Routledge, London, pp. 259-269.

Brusco, S. (1990), 'The idea of the Industrial District: Its genesis', in Pyke, F., Becattini, G. and Sengenberger, W. (eds), *Industrial districts and inter-firm co-operation in Italy*, International Institute for Labour Studies, Geneva, pp. 10-19.

Brusco, S. (1992), 'Small Firms and the Provision of Real Services', in Pyke, F. and Sengenberger, W. (eds), *Industrial Districts and Local Economic Regeneration*, International Institute for Labour Studies, Geneva, pp. 177-196.

Brusco, S. (1996), 'Trust, social capital and local development: some lessons from the experience of the Italian districts', in OECD (ed.), *Networks of Enterprises and Local Development: Competing and Co-operating in Local Productive Systems*, Paris, LEED for OECD, pp. 115-119.

Buckley, P.J. and Casson, M. (1985), *The Economic Theory of the Multinational Enterprise*, MacMillan, New York.

Burmeister, A. and Colletis-Wahl, K. (1997), 'Proximity in production networks: the circulatory dimension', *European Urban and Regional Studies*, vol. 4(3), pp. 231-241.

Burns, T. and Stalker, G. (1961), *The Management of Innovation*, Tavistock, London.

Burt, R.S. (1992), *Structural holes: The social Structure of Competition*, Harvard University Press, Cambridge MA.

Butera, F. (1990), *Il castello e la rete*, Milan , Angeli.

Bygrave, W.D. (1989), 'The Entrepreneurship Paradigm (I): A Philosophical Look at Its Research Methodologies', *Entrepreneurship Theory and Practice*, vol. 14(Fall), pp. 7-26.

Camagni, R. (ed.) (1991), *Innovation networks. Spatial perspectives*, Belhaven Press, London and New York.

Camagni, R. (1995), 'Global networks and local milieu: towards a theory of economi space', in Conti, S., Malecki E.J. and Oinas, P. (eds), *The Industrial Enterprise and Its Environment*, Aldershot, Avebury, pp. 195-214.

Carlsson, B. (1987), 'Reflection on Industrial Dynamics. The Challenges Ahead', *International Journal of Industrial Organization*, vol. 5(2), pp. 133-148.

Carter, J. R. and Narasimhan, R. (1996), 'A Comparison of North American and European Future Purchasing Trends', *International Journal of Purchasing and Materials Management*, May, pp. 12-22.

Castro, E. de and Jensen-Butler, C. (1993), *Flexibility, routine behaviour and the neo-classical model in the analysis of regional growth*, Department of Political Science, University of Aarhus, Denmark.

Cavusgil, S.T. (1984), 'Organizational characteristics associated with export activity', *Journal of Management Studies*, vol. 21(1), pp. 3-22.

Cavusgil, S.T., and Nevin, J.R. (1981), 'Internal determinants of export marketing behavior: An empirical investigation', *Journal of Marketing Research*, vol. 18(1), pp. 114-119.

Chandler, A.D. Jr. (1977), *The Visible Hand*, Harvard University Press, Cambridge MA.

Chapman, K. (1991), *The International Petrochemical Industry: Evolution and Location*, Blackwell, Oxford.

Chemie-Produktion (1997), Farben wollen hoch hinaus, 3/97, pp. 40-44.

Christensen, P.R. (1988), 'Enterprise Flexibility and Regional Networks', In Judét, P. (ed.), *Industrial Flexibility and Work*, Cahiers IREP, Grenoble.

Christensen, P.R. (1991), 'The small and medium-sized exporter's squeeze: Empirical evidence and model reflections', *Entrepreneurship and Regional Development*, vol. 3, pp. 49-65.

Christensen, P.R. and Lindmark, L. (1993), 'In search of regional support in internationalization of small and medium-sized firms - A network perspective', in Lundqvist, L. and Persson, L.O. (eds), *Visions and Strategies in European Integration. A North European Perspective*, Springer-Verlag, Berlin, pp. 131-151.

Christensen, P.R. and Philipsen, K. (1997), *Evolutionary Perspectives on the Formation of Small Business Clusters - the Case of a Stainless Steel District in Denmark*, Paper presented at RENT XI, Manheim.

Christensen, P.R., Eskelinen, H., Forsström, B., Lindmark, L. and Vatne, E. (1990), 'Firms in network: Concepts, spatial impacts and policy implications', in Illeris, S. and Jackobsen, L. (eds), *Networks and Regional Development*, NordREFO/Akademisk Forlag, Copenhagen.

Christopherson, S. (1993) ' Market Rules and Territorial Outcomes: The Case of the United States', *International Journal of Urban and Regional Research*, vol. 17, pp. 274-295.

Chusho Kigyo Dan Chusho Kigyo Kenkyujo (1987), *Nihon keizai no hatten to chusho kigyo: sengo no ayumi to yakuwari*, Tokyo, Douyoukan.

Ciampi, F. (1994), *Squilibri e assetto finanziario nelle P.M.I. Finanziamenti e contributi della Comunità europea*, Studi e Informazioni, Quaderni 45, Banca Toscana, Firenze.

Clark, G. (1992a), ''Real' Regulation: the administrative state', *Environment and Planning A*, vol.24, pp. 615-627.

Clark, G. (1992b), 'Problematic status of corporate regulation in the United States: towards a new moral order', *Environment and Planning A*, vol. 24, pp. 705-725.

Clark, G. L. and Wrigley, N. (1997a), 'The spatial configuration of the firm and the management of sunk costs', *Economic Geography*, vol. 73, pp. 285-304.

Clark, G.L. and Wrigley, N. (1997b), 'Exit, the firm and sunk costs: reconceptualizing the corporate geography of disinvestment and plant closure', *Progress in Human Geography*, vol. 21, pp. 338-358.

Clarke, I. (1984), 'The chemicals industry and ICI: the forms and impact of global corporations in Australia', in Taylor, M. (ed.) *The Geography of Australian Corporate Power*, Croom Helm Australia, Sydney, pp. 125-155.

Clarke, I. (1985), *The Spatial Organisation of Multinational Corporations*, Croom Helm, London.

Clarke, I. (1986), 'Labour dynamics and plant centrality in multinatiuonal corporations', in Taylor, M. and Thrift, N. (eds) *Multinationals and the Restructuring of the World Economy*, Croom Helm, London, 21-48.

Clegg, S. (1989), *Frameworks of Power*, Sage, London.

Clegg, S. and Dunkerley, D. (1980), *Organisation, Class and* Control, Routledge and Kegan Paul, London.

Coase, R.H. (1937), 'The Nature of the Firm', *Economica*, vol. 4, pp. 386-405.

Coleman, J. S. (1988), 'Social capital in the creation of human capital', *American Journal of Sociology*, vol. 94, pp. 95-120.

Conti, S. and Enrietti, A. (1995), 'The Italian Automobile Industry and the Case of Fiat: One Country, One Company, One Market?' in Hudson, R. and Schamp, E. (eds), *Towards a New Map of Automobile Manufacturing in Europe? New Production Concepts and Spatial Restructuring*, Springer Verlag, Berlin.

Conti, S. and Julien, P.A. (eds) (1991), *Miti e realtà del modello italiano, Letture sull'economia periferica*, Pàtron, Bologna.

Conti, S., Malecki, E.J. and Oinas, P. (eds) (1995), *The Industrial Enterprise and Its Environment: Spatial Perspectives*, Aldershot , Avebury.

Cooke, P. (ed.) (1995), *The Rise of the Rustbelt*, London, University College London Press.

Cooke, P. (1998), 'Introduction: origins of the concept', in Braczyk, H.J., Cooke, P. and Heidenreich, M. (eds), *Regional Innovation Systems*, London and Bristol PA, UCL Press, pp. 2-25.

Coriat, B. (1992), 'The revitalization of mass production in the computer age', in Storper M. and Scott A. J. (eds), *Pathways to Industrialization and Regional Development*,Routledge, London, pp. 137-156.

Cornish, S.L. (1997), 'Product innovation and the spatial dynamics of market intelligence: does proximity to markets matter?', *Economic Geography*, vol. 73, pp. 143-165.

Courault, B.A. and Romani, C. (1992), 'A reexamination of the Italian model of flexible production from a comparative point of view', in Storper, M. and Scott. A. (eds), *Pathways to Industrialization and Regional Development*, Routledge, London, pp. 205-215.

Cowling, K. and Sugden, R. (1987), 'Market exchange and the concept of a transnational corporation', *British Review of Economic Issues*, vol. 9, pp. 57-68.

Czarniawska-Joerges, B. and Wolff, R. (1989), 'Leaders, managers, entrepreneurs on and off the organizational stage', *Organization Studies*, vol. 12(4), pp. 529-546.

Czinkota, M.R. and Johnston, W.J. (1983), 'Exporting: does sales volume make a difference?', *Journal of International Business*, spring/summer, pp. 147-153.

Daft, R. L., and Weick, K. E. (1984), 'Toward a Model of Organizations as Interpretation Systems', *Academy of Management Journal*, vol. 19(2), pp 284-295.

Dahl, R. (1957), 'The concept of power', *Behavioural Science*, vol. 2, pp. 201-215.

Dalum, B. and Williamsen, G. (1996), 'Are OECD Export Specialization Patterns "Sticky"?', *Danish Research Unit for Industrial Dynamics* (DRUID), Working Paper.

Dandridge, T. and Johannisson, B. (1996), 'Entrepreneurship and Self-Organising: Personal Networks in Spatial Systems of Small Firms', in Zineldin, M (ed.), *Strategic Relationship Management*, Almqvist & Wiksell International, Stockholm, pp. 219-238.

Danmarks Statistik (1997), 'Undersøgelse af industriens brug af underleverandører 1994', *Nyt fra Danmarks Statistik*, no. 235.

Davidsson, P. (1995), 'Culture, Structure and Regional Levels of Entrepreneurship', *Entrepreneurship and Regional Development*, vol. 7(1), pp. 41-62.

DAW (1995), *Lebensräume gestalten und erhalten. Die Caparol-Firmengruppe im Jubiläumsjahr 1995*. Deutsche Amphibolin-Werke von Robert Marjahn, Ober-Ramstadt.

Deetz, S. (1992), *Democracy in an Age of Corporate Colonization - developments in Communication and the Politics of Everyday Life*, New York State University Press, Albany NY.

Delphi Automotive Systems [www.delphiauto.com].

Diamantopoulos, A. (1988), 'Identifying differences between high- and low-involvment exporters', *International Marketing Review*, Summer, pp. 56-60.

Dicken, P. (1992), *Global Shift: The Internationalization of Economic Activity*, Second Edition, Paul Chapman Publishing Ltd, London.

Dicken, P. (1994), 'Global-local tensions: firms and states in the global space economy', *Economic Geography*, vol. 70, pp.101-128.

Dicken, P. (1998), *Global Shift: Transforming the World Economy*, Third Edition, Guilford, New York.

Dicken, P. and Thrift, N.J. (1992), 'The organization of production and the production of organization: why business enterprises matter in the study of geographical industrialization', *Transactions of the Institute of British Geographers, New Series*, vol. 17, pp. 279-291.

Dicken, P. et. al. (1997), 'Unpacking the global', in Lee, R. and Wills, J. (eds), *Geographies of Economies*, Arnold, London, pp. 147-157.

Dippold, G. (1994), *Deutsches Korbmuseum Michelau. Begleitbuch zur Dauerausstellung*, Deutsches Korbmuseum, Michelau.

Dore, R. (1986), *Flexible Rigidities: industrial policy and structural adjustment in the Japanese economy, 1970-80*, Stanford University Press.

Dosi, G. (1988), 'The nature of the innovation process.' in Dosi, G. et al (eds), *Technical Change and Economic Theory*, Pinter, London.

Dosi, G., Pavitt, K. and Soete, L. (1990), *The Economics of Technical Change and International Trade*, Harvester Wheatsheaf, London.

Dunford, M.F., Fernandes, A., Musyck, B., Sadowski, B., Cho, M. and Tsenkova, S. (1993), 'The organization of production and territory: Small firm systems', *International Journal of Urban and Regional Research*, vol. 17 (1), pp. 132-136.

Dunning, D. (1988a), *Explaining International Production*, Unwin Hyman, Boston.

Dunning, J. (1988b), 'The eclectic paradigm of international production: A restatement and some possible extensions', *Journal of International Business Studies*, vol. 19(1), pp.1-31.

Dyer, J H and Singh, H. (1998), 'The Relational View: Cooperative Strategy and Sources of Interorganizational Competitive Advantage', *The Academy of Management Review*, vol. 23(4), pp. 660-679.

Easton, G. (1992), 'Industrial networks: a review', in Axelsson, B. and Easton, G. (eds) (1992), *Industrial Networks: A New View of Reality*, London, Routledge, pp.3-28.

Edmunds, S. and Sarkis, K. (1986), 'Exports: A necessary ingredient in the growth of small business firms', *Journal of Small Business Management*, vol. 24(4), pp. 54-65.

Ellegård, K. (1983), 'Människa - Produktion: tidsbilder av ett produktionssystem' (Man - Production: Time Pictures of a Production System), *Meddelanden från Göteborgs Universitets Geografiska institutioner*, Serie B Nr. 72, Göteborg, Kulturgeografiska institutionen, Göteborgs Universitet.

326 The Networked Firm in a Global World

Elsässer, B. (1995), *Svensk Bilindustri - en framgångshistoria*, SNS Förlag. Stockholm.

Enright, M. (1995), 'Regional clusters and economic development: a research agenda', in Staber, U., Schaefer, B. and Sharma, B. (eds), *Business Networks: Prospects for Regional Development*, de Gruyter, Berlin, pp. 190-214.

Erb W.-D. (1990), *Anwendungsmöglichkeiten der linearen Diskriminanzanalyse in Geographie und Regionalwissenschaft*, Weltarchiv, Hamburg.

Eskelinen, H., Lautanen, T. and Forsström, B. (1994), 'The Internationalization of SMEs in Four Finnish Regions: Some Tentative Observations', in Lundqvist L. and Persson, L.O. (eds), *Northern Perspectives on European Integration*, NordREFO/Scandinavian University Press, Copenhagen, pp. 115-133.

Esser, J. and Hirsch, J. (1994), 'The crisis of Fordism and the dimensions of a 'post-Fordist' regional and urban structure', in Amin A. (ed.) *Post-Fordism*, Blackwell, Oxford, pp. 71-97.

European Commission (1994), 'Pan-European Forum on Subcontracting.' *Economic Importance of Subcontracting in the Community*, Working document on the national reports, Working Group 1, Directorate-general Enterprise Policy, Distributive Trades, Tourism and Cooperatives.

European Commision (1995), *Green Paper on Innovation*, Bulletin of the European Union. Supplement 5/95, Luxembourg.

European Commission (1996), 'A European Approach to Strategic Alliances.' in *Panorama of EU Industry 1996*, Eurostat, Luxembourg, pp. 31-40.

Evans, P. (1995), *Embedded Autonomy: states and industrial transformation*, Princeton University Press, Princeton NJ.

Fahrmeir, L. and Hamerle, A. (1984), *Multivariate statistische Verfahren*, De Gruyter, Berlin.

Feenstra, R.C. (1998), 'Integration of Trade and Disintegration of Production in the Global Economy', *Journal of Economic Perspective*, vol. 12(4), pp. 31-50.

Ferrão, J. and Vale, M. (1995), 'Multi-purpose Vehicles, a New Opportunity for the Periphery? Lessons from the Ford/VW Project (Portugal)', in Hudson, R. and Schamp, E. (eds), *Towards a New Map of Automobile Manufacturing in Europe*, Springer Verlag, Berlin.

Financial Times (1996a), 'FT Survey - Automotive Components', May 21.

Financial Times (1996b), 'Components for success', May 8.

Financial Times (1996c), 'A driving force to dominate the industry', June 1-2.

Financial Times (1997b), 'Carmakers' smart move', July 1, p. 12.

Financial Times (1998), 'FT Auto Components', February 23.

Förth, St. (1995), *Die Korbflechtereien am Obermain*. Diploma dissertation, University of Nürnberg, Lehrstuhl für Wirtschafts- und Sozialgeographie.

Frank, H. and Lueger, M. (1995), 'Reconstructing Development Processes. Conceptual Basis and Empirical Analysis of Setting up a Business', *International Studies of Management and Organization*, vol. 27(3), pp. 34-63.

Freeman, C. (1990), 'Technical innovation in the world chemical industry and changes of techno-economic paradigm', in Freeman C. and Soete L. (eds) *New Exploitations in the Economics of Technical Change*, Frances Pinter, London, pp. 74-91.

Freeman, C. and Perez, C. (1988), 'Structural crises of adjustment, business cycles and investment behaviour', in Dose, G., Freeman, C., Nelson, R., Silverberg, G. and Soete, L. (eds), *Technical Change and Economic Theory*, Frances Pinter, London, pp. 38-66.

Friedman, D. (1988), *The Misunderstood Miracle: Industrial Development and Political Change in Japan*, Cornell University Press, Ithaca.

Fujimoto, T. (1997), *The Japanese Automobile Supplier System: Framework, Facts and Reinterpretation*, Discussion Papers Series, University of Tokyo, Research Institute for the Japanese Economy, Faculty of Economics.

Gadde, L.E. and Håkansson, H. (1992), *Organizing Purchasing - A Network Perspective*, Paper precented at the 8th IMP conference, Lyon.

Galbraith, J. K. (1968), *Il nuovo stato industriale*, Einaudi, Turin (original edition, 1967).

Garofoli, G. (1991), 'The Italian model of spatial development in the 1970s and 1980s', in Benko, G. and Dunford, G.M. (eds), *Industrial Change and Regional Development*, Belhaven Press, London.

Gartner, W.B., Bird, B.J and Starr, J.A. (1992), 'Acting 'As If': Differentiating Entrepreneurial from Organizational Behavior', *Entrepreneurship Theory and Practice*, Spring, pp. 13-31.

Geertz, C. (1983), *Local Knowledge: Further Essays in Interpretative Anthropology*, Basic Books, New York.

Gerlach, M.L. (1992), *Alliance Capitalism: The Social Organization of Japanese Business*, University of California Press, Berkeley.

Gerschenkron, A. (1962), *Economic Backwardness in Historical Perspective: A book of essays*, The Belknap Press of Harvard University Press, Cambridge MA.

Gertler, M.S. (1992), 'Flexibility revisited: districts, nation-states, and the forces of production', *Transactions of the Institute of British Geographers NS*, vol. 17, pp. 259-278.

Gertler, M.S. (1993), 'Implementing advanced manufacturing technologies in mature industrial regions: towards a social model of technology production', *Regional Studies*, vol. 27, pp. 665-680.

Gertler, M.S. (1996), 'Worlds apart: the changing market geography of the German machinery industry?', *Small Business Economics*, vol. 8, pp. 87-106.

Giddens, A. (1984), *The Constitution of Society*, Polity Press, Oxford.

Gilly, J. P. (1994), 'Dinamiche Industriali e meso-analisi. Il caso dei sistemi di innovazione', *L'industria*, vol. 2, pp. 295-309.

Glasmeier, A.K. (1991), 'Technological discontinuities and flexible production networks: the case of the world watch industry', *Research Policy*, vol. 20, pp. 469-485.

Glasmeier, A. K. (1999), 'Globalization/marginalization: two sides of the same coin?', in Stokke, K. and Fraas. M. (eds), *Globalisering. Rapport for Norske geografers forenings årskonferanse 1999*. Occasional Paper no. 26, Human Geography, Department for Sociology and Human Geography, University of Oslo.

Glasmeier, A. K. and McCluskey, R.E. (1987), 'U.S. Auto Parts Production: An Analysis of the Organization and Location of a Changing Industry', *Economic Geography*, vol. 63, pp. 142-159.

Glasmeier, A.K. and Sugiura, N. (1991), 'Japan's Manufacturing System: SME, Subcontracting and Regional Complex Formation', *International Journal of Urban and Regional Research*, vol. 15, pp. 395-414.

Gold, T.B. (1989), *State and Society in the Taiwan Miracle*, Sharpe, Armonk NY.

Goldenberg, S. and Haines, V.A. (1992), 'Social networks and institutional completeness: from territory to ties', *Canadian Journal of Sociology*, vol. 3, pp. 301-12.

Goodman, E. (1989), 'Introduction: the political economy of the small firm in Italy.' in Goodman, E. and Bamford, J. (eds), *Small Firms and Industrial Districts in Italy*, Routledge, London.

Grabher, G. (1993a), 'Rediscovering the social in the economics of interfirm relations', in Grabher, G. (ed.), *The Embedded Firm: On the Socioeconomics of Industrial Networks*, Routledge, London, pp. 1-31.

Grabher, G. (ed.) (1993b), *The Embedded Firm: On the Socioeconomics of Industrial Networks*, Routledge, London.

Grabher, G. (1993c), 'The Weakness of Strong Ties: The Lock-In of Regional Development in the Ruhr Area', in Grabher, G. (ed.), *The Embedded Firm. On the Socioeconomics of Industrial Networks*, Routledge, London, pp. 255-277.

Granovetter, M. (1985), 'Economic action and social structure: the problems of embeddedness', *American Journal of Sociology*, vol. 91(3), pp. 481-510.

Granovetter, N. and Swedberg, R. (1985), *The Sociology of Economic Life*, Westview Press, Boulder, Colorado.

Green, R. and Larsen, T. (1987), 'Environmental shock and export opportunity', *International Marketing Review*, vol. 4(4), pp. 30-42.

Gregersen, B. and Johnson, B. (1997), 'Learning Economies, Innovation Systems and European Integration', *Regional Studies*, vol. 31, pp. 479-490.

Haffert, K. (1957), *Die industrielle Standortdynamik im Raum von Coburg nach dem Zweiten Weltkrieg*, PhD dissertation, University of Erlangen-Nürnberg.

Haid, A. and Wessels, H. (1997), 'Schrumpfungsprozeá der deutschen Schuhproduktion halt an', *DIW-Wochenbericht*, 14/97.

Håkansson, H. and Snehota, I. (1995), *Developing Relationships in Business Networks*, Routledge, London.

Håkansson, H. and Johanson, J. (1993), 'The network as a governance structure: interfirm cooperation beyond markets and hierarchies', in Grabher, G., (ed.), *The Embedded Firm: On the Socioeconomics of Industrial Networks*, London, Routledge, pp. 35-51.

Hallsworth, A.G. and Taylor, M.J. (1996), 'Buying power: interpreting retail change in a circuits of power framework', *Environment and Planning A*, vol. 28, pp. 2125-2137.

Hanack, P. (1995), Der Lack läßt den Sportwagen zu Stein erstarren, *Frankfurter Rundschau*, 28 January, p. 13.

Harrison, B. (1992), 'Industrial Districts: Old Wine in New Bottles?' *Regional Studies*, vol. 25, pp. 469-483.

Harrison, B. (1994a), 'The Myth of Small Firms as the Predominant Job Generators', *Economic Development Quarterly*, vol. 8(1), pp. 3-18.

Harrison, B. (1994b), *Lean and mean: the changing landscape of corporate power in the age of flexibility*, Basic Books, New York.

Harrison, B. (1997), *Lean and Mean. The Changing Landscape of Corporate Power in the Age of Flexibility*, Second Edition, Guilford, New York.

Harvey, D. (1990), *The Condition of Postmodernity. An Enquiry Into the Origins of Cultural Change*, Blackwell, Cambridge, MA.

Hayter, R. (1997), *The Dynamics of Industrial Location: The Factory, the Firm and the Production System*, John Wiley and Sons, Chichester.

Hedlund, G. and Rolander, D. (1990), 'Actions in Hierarchies: New Approaches to Managing the MNCs', in Bartlett, C., Doz, Y. and Hedlund, G. (eds), *Managing the Global Firm*, Routledge, London.

Heidenreich, M. and Krauss, G. (1998), 'The Baden-Württemberg production and innovation regime: past successes and new challenges', in Braczyk, H.J., Cooke, P. and Heidenreich, M. (eds), *Regional Innovation Systems*, UCL Press, London and Bristol PA, pp. 214-244.

Helper, S. (1991), 'How Much Has Really Changed between U.S. Automakers and Their Suppliers?', *Sloan Management Review*, Summer, pp. 15-28.

Helper, S. (1993), 'An exit-voice analyses of supplier relations' in Grabher, G. (ed.), *The Embedded Firm: On the Socioeconomics of Industrial Networks*, Routledge, London, pp. 141-160.

Helper, S. and Sako, M. (1995), 'Supplier Relations in Japan and the United States: Are They Converging?', *Sloan Management Review*, Spring, pp. 77-84.

Henneking, R. (1994), *Chemische Industrie und Umwelt*, Steiner, Stuttgart.

Herrigel, G. (1993), 'Power and the redefinition of industrial districts: the case of Baden- Württemberg', in Grabher G. (ed.), *The Embedded Firm: On the Socioeconomics of Industrial Networks*, Routledge, London, pp. 227-252.

Hervik, A. et al (1998), *Utviklingen i maritime næringer i Møre og Romsdal*. Rapport nr. 98/05, Molde, Møreforskning.

Hill, R. C. (1989). 'Comparing transnational production systems: the automobile industry in the USA and Japan.' *International Journal of Urban and Regional Research*, 13(3), pp. 462-479.

Hines, P. (1994), *Creating World Class Suppliers - Unlocking Mutual Competitive Advantage*, Pitman Publishing, London.

Hirsch, J. (1990), *Kapitalismus ohne Alternative?*, VSA, Hamburg.

Hirschman, A.O. (1958), *The Strategy of Economic Development*, Yale University Press, Newhaven.

Hjorth, D. and Johannisson, B. (1997), *The Ugly Duckling of Organising - on Entrepreneurialism and Managerialism*, Paper presented at the 42 ICSB Conference, San Francisco, California, June.

Hofmann, J. (1998), Für einen Teil der deutchen Fabriken lauft die Zeit ab. *Handelsblatt*, no. 144, 30.7.1998, p. 12.

Holmes, J. (1986), 'The organization and localization of production subcontracting', in Scott, A.J. and Storper, M.J. (eds), *Production, Work, Territory. The Geographical Anatomy of Industrial Capitalism*, Allen and Unwin, Boston MA.

Holton, R. J. (1992), *Economy and Society*, Routledge, London.

Hoppenstedt (1993a), *Mittelständische Unternehmen 1993, Band 1: Alphabetisches Firmenregister, Firmenberichte der Orte A-J*, Eighth Edition, Hoppenstedt, Darmstadt.

Hoppenstedt (1993b), *Mittelständische Unternehmen 1993, Band 2: Firmenberichte der Orte K-Z, Branchenregister, Ortsregister*, Eighth Edition, Hoppenstedt, Darmstadt.

Hoppenstedt (1993c), *Handbuch der Großunternehmen 1993, Band 1: Firmenberichte der Orte A-J, Alphabetisches Firmenregister*, Fortieth Edition, Hoppenstedt, Darmstadt.

Hoppenstedt (1993d), *Handbuch der Großunternehmen 1993, Band 2: Firmenberichte der Orte K-Z, Branchenregister, Ortsregister*, Fortieth Edition, Hoppenstedt, Darmstadt.

Hovers, A. and Wever, E. (1989), 'The alert firm: spatial patterns and theory', in de Smidt, M. and Wever, E. (eds), *Regional and Local Economic Policies and Technology*, Netherlands Geographical Studies, No. 99, pp. 131-143.

Howard J.P. (1996) 'Trends in the world paint industry', *Chemistry & Industry*, vol. 21, October, pp. 796-799.

Hoy, F. and Verser, T.G. (1994), 'Emerging Business, Emerging Field: Entrepreneurship and the Family Firm', *Entrepreneurship Theory & Practice*, vol. 18 (Fall), pp. 9-23.

Hudson, R. (1999), 'The learning economy, the learning firm and the learning region: a sympathetic critique of the limits to learning', *European Urban and Regional Studies*, vol. 6, pp. 59-72.

Hudson, R. and Schamp, E. (eds) (1995), *Towards a New Map of Automobile Manufacturing in Europe? New Production Concepts and Spatial Restructuring*, Springer Verlag, Berlin.

Hull, C.J. and Hjern, B. (1987), *Helping Small Firms Grow. An implementation approach*, Croom Helm, London.

Imai, K. and Baba, Y. (1991), 'Systemic innovation and cross border networks: transcending markets and hierarchies to create a new techno-economic system', in *Technology and Productivity*, OECD, Paris, pp. 389-407.

Imrie, R. and Morris, J. (1992), 'A Review of Recent Changes in Buyer–Supplier Relations', *Omega*, vol. 20(5/6), pp. 641-652.

Indergaard, M. (1996), 'Making networks, remaking the city', *Economic Development Quarterly*, vol. 10, pp. 172-187.

Isaksen, A. (1990), 'Spatial division of labour in Norway: dynamic centre, traditional periphery', *Norsk Geografisk Tidsskrift*, vol. 44, pp. 53-60.

Isaksen, A. (1994), *Regional næringsutvikling og fremvekst av spesialiserte produksjonsområder*, Agderforsknings Skriftserie, No.5, Kristianssand.

Isaksen, A. et al (eds) (1999), *SME Policy and the Regional Dimension of Innovation. The Norwegian Report*, SMEPOL Report No. 5. STEP Group, Oslo.

Itoh, Y. (1958), *Regional Paradigms for the Japanese Textile Industry: Case of Ichinomiya*, (in Japanese), Jinbun Chiri (Human Geography).

Japan Center for Economic Research, (1999), *Nihon ni okeru bencha kigyo no genjo to kadai*. [www.jcer.or.jp/jpn/b/keizai/98sangyo.htm].

Japan External Trade Organization (1989), *SME Law and Public Policy Analysis*, Market Research Group, Inc., New York.

Japan Small Business Research Institute (1995), *White Paper on Small and Medium Enterprises in Japan*, MITI, Tokyo.

Japan Small Business Research Institute (1998a), *A Study on the Present Conditions and the Future Outlook of Industrial Districts in Japan. Structures of New Linkages among Companies*, JSBRI, Tokyo.

Japan Small Business Research Institute (1998b), *White Paper on Small and Medium Enterprises in Japan - The Need for Small and Medium Enterprises to Change and Display Entrepreneurship*, MITI, Tokyo.

Jessop, B. (1990), 'Regulation theories in retrospect and prospect', *Economy and Society*, vol. 19, pp.153-216.

Jessop, B. (1992), 'Fordism and post-Fordism: a critical reformulation.' in Storper M.J. and Scott A.J. (eds), *Pathways to Industrialization and Regional Development*, Routledge, London, pp. 46-69.

Jessop, B. (1993), 'Towards a Schumpeterian workfare state? Preliminary remarks on post-Fordist political economy', *Studies in Political Economy*, vol. 40, pp. 7-39.

Johannisson, B. (1987), 'Toward a Theory of Local Entrepreneurship', in Wyckham, R.G. and Meredith, L.N. and G.R. (eds), *The Spirit of*

Entrepreneurship, Simon Fraser University, Faculty of Business Administration. Vancouver BC, pp. 1-14.

Johannisson, B. (1988), 'Business Formation - A Network Approach', *Scandinavian Journal of Management*, vol. 4(3/4), pp. 83-99.

Johannisson, B. (1990), 'Community entrepreneurship: cases and conceptualization', *Entrepreneurship and Regional Development*, vol. 2, pp. 71-88.

Johannisson, B. (1992), 'Entrepreneurship - The Management of Ambiguity', in Polesie, T. and Johansson, I.L. (eds), *Responsibility and Accounting. The Organizational Regulation of Boundary Conditions*, Studentlitteratur, Lund, pp. 155-179.

Johannisson, B. (1996), 'Existential Enterprise and Economic Endeavour', *Aspects of Women's Entrepreneurship*. Nutek B 1996, No.10. Stockholm, pp. 115-141.

Johannisson, B., Alexanderson, O., Nowicki, K. and Senneseth, K. (1994), 'Beyond Anarchy and Organization - Entrepreneurs in Contextual Networks', *Entrepreneurship and Regional Development*, vol. 6, pp. 329-356.

Johansson, J. and Mattson, L.G. (1986), 'Interorganizational relations in industrial systems: a network approach compared with the transaction cost approach', *Working Paper 1987/7*, Department of Business Administration, University of Uppsala.

Johansson J. and Mattsson, L.G. (1987), 'Interorganizational relations in industrial systems: a network approach compared with thetransaction cost approach', *International Studies of Management Organization*, vol. 17(1), pp. 34-48.

Johansson, J. and Mattsson, L.G. (1991), 'Interorganisational relations in industrial systems: a network approach compared with the transactions-cost approach', in Thompson, G., Frances, J. et al, (eds), *Markets, Hierarchies and Networks: The Coordination of Social Life*, Sage Publications, London.

Johnson Controls [www.johnsoncontrols.com].

Jones, D. (1990), *Beyond the Toyota Production System: The Era of Lean Production*, Fifth International Operations Management Association Conference on Manufacturing Strategy, Warwick, pp. 1-13.

Jorde, M. T. and Teece, D. J. (1989), 'Competition and Cooperation: Striking the Right Balance', *California Management Review*, Spring, pp. 25-37.

Julien, P.A. (1995), *Mondialisation des marchés et type de comportements de PME*, Cahiers de recherche - Chaire Bombardier, Université du Québec à Trois-Rivières.

Julien, P.A. and Marchesnay, M. (1988), *La Petite Entreprise*, Vuibert, Paris.

Julien, P.A. and Maurel, B. (1986), *La Belle Entreprise*, Boréal, Québec.

Kalsaas, B.T. (1998), 'Paths of Development in the Japanese Automotive Industry. Changing Competitiveness and the Just-in-Time System', in van Grunsven, L. (ed.), *Regional Change in Industrializing Asia*, Ashgate, Aldershot.

Kaynak, E., Ghauri, P.N. and Olofsson-Bredenlöw, T. (1987), 'Export behavior of small Swedish firms', *Journal of Small Business Management*, vol. 25(2), pp. 26-32.

Kearney, A.T. Consulting (1993), *Logistics Excellence in Europe*, European Logistics Association.

Keeble, D. and Wilkinson, F. (1999), 'Collective Learning and Knowledge Development in the Evolution of Regional Clusters of High Technology SMEs in Europe', *Regional Studies*, vol. 33(4), pp. 95-303.

Keiei Joho Shuppan (1988), *Japan's SMEs Today: A Closer Look at 100 Industrial Segments*, Keiei Joho Shuppan, Tokyo.

Kennedy, M. and Florida, R. (1993), *Beyond Mass Production: The Japanese System and its Transfer to the US*, Oxford University Press, Oxford.

Kenney, M. and Von Burg, U. (1999), 'Technology, Entrepreneurship and Path Dependence: Industrial Clustering in Silicon Valley and Route 128', *Industrial and Corporate Change*, vol. 8(1), pp. 67-103.

Kern, H. and Schumann, M. (1990), *Das Ende der Arbeitsteilung? Rationalisierung in der industriellen Produktion*, Fourth Edition, C.H. Beck, Munich.

Knoke, D. (1990), *Political Networks: The Structural Perspective*, Cambridge University Press, New York.

Koschatzky, K., Hemer, J. and Gundrum, U. (1996), *Verflechtungsbeziehungen von Produktions- und Dienstleistungsunternehmen in der Region Rhein Main*. Fraunhofer Institut für Systemtechnik und Innovationsforschung, Karlsruhe.

Krugman P. (1991) *Geography and Trade*, Leuven University Press, Leuven & MIT Press, Cambridge, MA.

Krugman, P. (1998), *Japan's Trap. A Monograph*. May. personal Web-site.

Kuhligk, S. (1976), *Standortliche Analyse der monoindustriellen Agglomeration. Dargestellt am beispiel des aktuellen monoindustriellen Pirmasenser Problemgebietes unter besonderer Berücksichtigung regionalpolitischer Aspekte*. Diss. Free University Berlin, Berlin.

Kurose, N. (1997), *Chusho kigyo seisaku no soukatsu to teigen*, Doyukan, Tokyo.

Laigle, L. (1997), *The Local Embedding of International Supplier Plants*, Paper presented at the EUNIT International Conference on Industry, Innovation and Territory. 20 - 22 March 1997, Lisbon, Portugal.

Lamming, R. (1993), *Beyond Partnership – Strategies for Innovation and Lean Supply*, Prentice Hall, New York.

Larsson, A. (1993), Underleverantörssystem i förändring: geografiska aspekter på 'just-in-time' tillverkning. (Changing subcontracting systems: geographical aspects on 'just-in-time' manufacturing), Licentiat-uppsats (Masters-thesis), *CHOROS 1993:4*, Göteborg, Handelshögskolan vid Göteborgs Universitet, Kulturgeografiska institutionen.

Laufer, J. (1975), 'Comment on devient entrepreneur?' *Revue Française de Gestion*, vol. 2, pp. 13-15.

Lawrence, P and Lorsch, J. (1967), *Organization and Environment: Managing Differentiation and Integration*, Graduate School of Business administration, Harvard.

Lawson, C. and Lorenz, E. (1999), 'Collective Learning, Tacit Knowledge and Regional Innovative Capacity', *Regional Studies*, vol. 33(4), pp. 305-317.

Lear Corporation [www.lear.com].

Leborgne, D. and Lipietz, A. (1992), 'Conceptual fallacies and open questions on post-Fordism', in Storper, M. and Scott, A. (eds), *Pathways to Industrialization and Regional Development*, Routledge, London, pp. 332-348.

Lecoq, B. (1993), 'Proximité et rationalité économique,' *Revue d'économie Régionale et Urbaine*, vol. 3, pp. 469-486.

Lipietz, A. (1987), *Mirages and Miracles. The Crises of Global Fordism*, Verso, London.

Lorenzen, M. (ed.) (1998), *Specialisation and Localised Learning*, Copenhagen Business School, Copenhagen.

Lovering, (1999), 'Theory Led by Policy: The Inadequacies of the 'New Regionalism'', *International Journal of Urban and Regional Research*, vol. 23, pp. 379-395.

Lundberg, L. (1992), 'European Economic Integration and the Nordic countries trade', *Journal of Common Market Studies*, vol. 302), pp. 157-173.

Lundvall, B.-Å. (1988), 'Product innovation and user-producer relations', in Dosi, G. et al. (eds), *Technical Change and Economic Theory*, Pinter, London.

Lundvall, B.-Å. (ed.) (1992), *National Systems of Innovation*, Pinter, London.

Lundvall, B.-Å. (1993), 'Explaining Interfirm Cooperation and Innovation: Limits of the Transaction-Cost Approach', in Grabher, G (ed.), *The Embedded Firm. On the Socioeconomics of Industrial Networks*, Routledge, London, pp. 52-64.

Lundvall, B.-Å. (1996), 'The Social Dimension of The Learning Economy', *DRUID Working Paper* No. 96-1, Department of Business Studies, Aalborg University.

Lundvall, B.-Å . (1998), 'Defining Industrial Dynamics and its Research Agenda', *Paper to the DRUID Winter Seminar*, Middelfart.

Lundvall, B.-Å. and Borrás, S. (1997), 'The globalising learning economy: Implications for innovation policy', *Report from DG XII*, Commission of the European Union.

Lundvall, B.-Å. and Johnson, B. (1994), 'The learning economy', *Journal of International Studies*, vol. 1(2), pp. 23-42.

Lüthje, T. (1999), 'Halvfabrikatas betydning for dansk økonomi', *Økonomi & Politik*, vol. 1, pp. 41-53.

MacPherson, A. D. (1992), 'Innovation, External Technical Linkages and Small Firm Commercial Performance: an Empirical Analysis from Western New York', *Entrepreneurship & Regional Development*, vol. 4(2), pp. 165-184.

Maillat, D. (1998), 'Interactions between urban systems and localised productive systems: an approach to endogenous regional development in terms of innovative milieu', *European Planning Studies*, vol. 6(2), pp. 117-129.

Maillat, D., Crevoisier, O. and Lecoq, B. (eds) (1991), *Réseaux d'innovation et milieux innovateurs: un pari pour le développement régional*, Edes, Neuchatel.

Mair, A. (1994), *Honda's Global Local Corporation*, St. Martin's Press, New York.

Mair, A. and Florida, R. et al (1988), 'The new geography of automobile production: Japanese transplants in North-America', *Economic Geography*, vol. 64, pp. 352-373.

Malmberg, A. (1997), 'Industrial geography: location and learning', *Progress in Human Geography*, vol. 21(4), pp. 573-582.

Malmberg, A. and Maskell, P. (1997), 'Towards and explanation of regional specialization and industry agglomeration', in Eskelinen, H. (ed.), *Regional Specialisation and Local Environment - Learning and Competiveness*, NordREFO/NORDREGIO, Copenhagen.

Malmberg, A., Sölvell, Ö. and Zander, I. (1996), 'Spatial clustering, local accumulation of knowledge and firm competitiveness', *Geografiska Annaler*, vol. 78 B(2), pp. 85-97.

Markusen, A. (1996), 'Sticky places in slippery space: a typology of industrial districts', *economic Geography*, vol. 72(3), pp. 293-313.

Marsden, P. (1992), 'Real regulation reconsidered.' *Environment and Planning, A*, vol. 24(7), pp. 51-67.

Marshall A. (1927), *Industry and Trade. A Study of Industrial Technique and Business Organization; and Their Influences on the Conditions of Various Classes and Nations*, Third Edition, Macmillan, London.

Marshall, A. (1896), *Principles of Economics*, Second Edition, MacMillan, London.

Marshall, A. 1920 (1979), *Principles of Economics*, Eighth Edition, MacMillan, London.

Martinelli, F. and Schoenberger, E. (1992), 'Oligopoly is alive and well: Notes for a broader discussion of flexible accumulation', in Benko, G. and Dunford, M. (eds), *Industrial Change and Regional Development: The Transformation of New Industrial Spaces*, Belhaven Press, London, pp. 117-133.

Maskell, P. (1998), 'Low-tech competitive advantages and the role of proximity: the Danish wooden furniture industry', *European Urban and Regional Studies*, vol. 5, pp. 99-118.

Maskell, P. (1999), 'Social capital, innovation and competitiveness', unpublished manuscript.

Maskell, P. and Malmberg, A. (1997), 'Towards an Explanation of Regional Specialization and Industrial Agglomeration', *European Planning Studies*, vol. 5, pp. 25-41.

Maskell, P. and Malmberg, A. (1998), *Explaining the Location of Economic Activity: 'Ubiquitification' and the Importance of Learning*, Paper presented at the 94th Annual Meeting of the Association of American Geographers, 25-29 March, Boston.

Maskell, P. and Malmberg, A. (1999), 'Localised learning and industrial competitiveness', *Cambridge Journal of Economics*, vol. 23, pp. 167-190.

Maskell, P., Eskelinen, H., Hannibalsson, I., Malmberg, A. and Vatne, E. (1998), *Competitiveness, Localised Learning and Regional Development. Specialisation and Prosperity in Small Open Economies*, Routledge, London.

Massey, D. (1997), 'Economic/Non-economic', in Lee, R. and Wills, J. (eds), *Geographies of Economies*, Arnold, London, pp. 27-37.

Miles, M. B. and Huberman, A. M. (1994), *Qualitative Data Analysis. An Expanded Sourcebook*, Second Edition, Sage, Thousand Oaks, CA.

Miller, E.J. and Rice, A.K. (1967), *Systems of Organization*, Tavistock, London.

Mindlin, S. and Aldrich, H. (1975), 'Interorganizational dependence: a review of the concept and a re-examination of the findings of the Aston Group', *Administrative Science Quarterly*, vol. 20, pp. 382-392.

Ministry of International Trade and Industry (1989a), *Outline of the Small and Medium Enterprise Policies of the Japanese Government*, Small and Medium Enterprise Agency, Japan SME Corporation.

Ministry of International Trade and Industry (1989b), *SME in Japan 1989: White Paper on Small and Medium Enterprises in Japan*, Small and Medium Enterprise Agency.

Miwa, Y. (1990), *Nihon no kigyo to sangyo soshiki*. Tokyo, Daigaku Shuppankai.

Miyakawa, K. (1964), 'The dual structure of the Japanese economy and the growth pattern', *Developing Economies*, vol. 2, pp. 147-170.

Miyashita, K. and Russel, D. W. (1994), *Keiretsu: Inside the hidden Japanese conglomerated*, McGraw-Hill, New York.

Morgan, G. (1993), *Imaginization*, Sage, Newbury Park CA.

Morgan, K. (1997), 'The learning region: institutions, innovation and regional renewal', *Regional Studies*, vol. 31(5), pp. 491-504.

Moulaert, F. and Swyngedouw, E.A. (1989), 'A regulation approach to geography of flexible production systems', *Environment and Planning D: Society and Space*, vol. 7(3), pp. 327-345.

Müller-Fürstenberger, G. (1995), *Kuppelproduktion. Eine theoretische und empirische Analyse am Beispiel der chemischen Industrie*, Physica, Heidelberg.

Nadvi, K. and Schmitz, H. (1994), *Industrial Clusters in Less Developed Countries: Review of Experiences and Research Agenda*, IDS Discussion Papers 339, Brighton.

Nakamura, T. (1983), *Chusho kigyo to daikigyou*, Tokyo Keizai Shinposha, Tokyo.

Nås, S.O. (1998), 'Innovasjon i Norge – en statusrapport', *STEP Report R-08-1998*, STEP Group, Oslo.

Nihon Keizai Shimbun (1998a), Kigyoka no tamago wo ikusei: jichitai aitsugi josei seido, 27 April.
Nihon Keizai Shimbun (1998b), Daigakusei ni kigyo kyoiku, 21April.
Nihon Keizai Shimbun (1998c), VC no toushi sonshitsugaku zouka, 10 March.
Nihon Keizai Shimbun (1998d), Chiiki VB ikusei ni shiren: jichitai suisho kigyo airzugi tousan, 30 March.
Nihon Keizai Shimbun (1998e), Kigyo saisei 1: oikaze ga yande, 20 May.
Nishiguchi, T. (1993), *Strategic Industrial Sourcing,* Oxford University Press, New York.
Nishiguchi, T. and Brookfield, J. (1997), 'The Evolution of Japanese Subcontracting', *Sloan Management Review,* Fall, pp. 89-101.
Nonaka, I. and Takeuchi, H. (1995), *The Knowledge-Creating Company,* Oxford University Press, Oxford.
Norsk Hydro ASA [www.hydro.com].
North, D.C. (1990), *Institutions, Institutional Change and Economic Performance,* Cambridge University Press, Cambridge.
Nygaard, C. (1998), *Towards an Understanding of the Effect of Institutionalised Codes of Conduct of Industrial Organisation,* Paper presented at the 14th EGOS Colloquium in Maastricht, July.
OECD (1992), *Globalization of Industrial Activities,* OECD, Paris.
OECD (1996a), *SMEs: Employment, Innovation and Growth: The Washington Workshop,* OECD, Paris.
OECD (1996b), *OECD Letter 5/8.*
OECD (1998), *Issues Paper,* International Conference on 'Building Competitive Regional Economies: Up-grading Knowledge and Diffusing Technology to Small Firms.' Modena, Italia, 28-29 May 1998.
Office of Local Government, (1992), *Regional Economic Strength, Resilience and Vulnerability,* OLG Working Paper, Canberra, Australia.
Oinas, P. (1995), 'On the socio-spatial embeddedness of business firms', *Erdkunde,* vol. 51, pp. 23-32.
O'Rourke, D.A. (1985), 'Differences in exporting practices, attitudes, and problems by size of firm', *American Journal of Small Business,* Winter, pp. 68-73.
Ouchi, W.G. (1980), 'Markets, Bureaucracies and Clans', *Administrative Science Quarterly,* vol. 25, pp. 129-141.
Park, S.O. (1996), 'Networks and embeddedness in the dynamic types of new industrial districts', *Progress in Human Geography,* vol. 20, pp. 476-493.
Patchell, J., Hayter, R. and Rees, K. (1999), 'Innovation and local development: the negelected role of large firms', in Malecki, E. and Oinas, P. (eds) *Making Connections: Technological Learning and Regional economic Change,* Ashgate, Aldershot, pp. 109-142.
Penrose, E. (1959), *The Theory of the Growth of the Firm,* Basil Blackwell, Oxford.

Perrow, C. (1992), 'Small-Firm Networks', in Nohria, N. and Eccles, R.G. (eds), *Networks and Organizations*, Harvard Business School, Boston MA.

Pfeffer J. (1981), *Power in Organizations*, Pitman, Marshfield, MA.

Pfeffer, J. and Salancik, G.R. (1978), *The External Control of Organizations*, Harper & Row, New York.

Pike, A. and Tomaney, J. (1999), 'The Limits to Localization in Declining Industrial Regions? Trans-National Corporations and Economic Development in Sedgefield Borough', *European Planning Studies*, vol. 7(4), pp. 407-428.

Piore, M.J. and Sabel, C.F. (1984), *The Second Industrial Divide: possibilities for prosperity*, Basic Books, New York.

Polanyi, M. (1967), *The Tacit Dimension*, Routledge and Kegan, London.

Porter, M. (1985), *Competitive Advantage*, Free Press, New York.

Porter, M. (1990), *The competitive advantage of nations*, Macmillan, London.

Porter, M. (1998), 'Clusters and the new economics of competition', *Harvard Business Review*, November-December, pp. 77-90.

Powell, W.W. (1990), 'Neither markets nor hierarchies: network forms of organization', *Research in Organizational Behavior*, vol. 12, pp. 295-336.

Powell, W.W. and Smith-Doerr, L. (1994), 'Networks and economic life', in *The Handbook of Economic Sociology*, Princeton University Press, New York.

PR Newswire, Company News on the Internet [www.prnewswire.com], Borgers GmbH [www.borgers.de].

Prahalad, C.K. and Bettis, R.A. (1986), 'The Dominant Logic: a New Linkage between Diversity and Performance', *Strategic Management Journal*, vol. 7, pp. 485-501.

Pratt, A. (1997), 'The emerging shape and form of innovation networks and institutions', in Simmie, J. (ed.), *Innovation, Networks and Learning Regions?*, Regional Policy and Development Series 18, Jessica Kingsley Publishers and Regional Studies Association, London and Bristol PA, pp. 124-36.

Prop. 1997/98:62 *Regional tillväxt - för arbete och välfärd*, Stockholm.

Pulignano, V. (1997), *The Structure of the Supply Chain in the 'Integrated Factory' Model. The Case of FIAT Car Plant in Melfi*, University of Calabria, Department of Sociology and Political Sciences.

Putnam, R. D. (1993), 'The prosperous community. Social capital and public life', *The American Prospect*, vol. 13, pp. 35-42.

Putnam, R.D. (1993), *Making Democracy Work: Civic Traditions in Modern Italy*. Princeton University Press, Princeton NJ.

Pyke, F., Beccatini, G. and Sengenberger, W. (1992), *Industrial Districts and Interfirm Co-operation in Italy*, International Institute for Labour Studies, Geneva.

Rabelotti, R. (1995), 'Is There an "Industrial District Model"? Footwear Districts in Italy and Mexico Compared', *World Development*, vol. 23, pp. 29-41.

Rainnie, A. (1989), *Industrial Relations in Small Firms. Small Isn't Beautiful*, Routledge, London.

Rainnie, A. (1993), 'The reorganization of large firms subcontracting: myth and reality.' *Capital & Class,* vol. 49, pp. 53-75.

Rehle, N. (1996), 'Zwischen Moderne und Tradition. Essay zur Zukunft der Gablonzer Industrie', in Kulturamt der Stadt Kaufbeuren (ed.), *1946-1996. 50 Jahre Neugablonz,* Kulturamt der Stadt, Kaufbeuren, pp.45-71.

Reid, S. (1984), 'Information acquisition and export entry decisions in small firms', *Journal of Business Research,* vol. 12(2), pp. 141-157.

Reve, T. (1990), 'The firm as a nexus of internal and external contracts', in Aoki, M., Gustafson, B. and Williamson O.E. (eds), *The Firm as a Nexus of Treaties,* Sage, London.

Rieter Group [www.rieter.com].

Ring, P.S. and Van de Ven, A. (1994), 'Developmental Processes of Interorganizational Relationships', *Academy of Management Review,* vol. 19(1), pp. 90-118.

Rosenfeld, S. (1992), *Competitive Manufacturing: New Strategies for Regional Development,* Center for Urban Policy Research, Rutgers University, New Brunswick, NJ.

Rosenfeld, S. (1997), 'Bringing Business Clusters into the Mainstream of Economic Development', *European Planning Studies,* vol. 5(1), pp. 3-23.

Rostow, W.W. (1952), *The Process of Economic Growth,* Norton, New York.

Rothwell, R. and Zegveld, W. (1982), *Innovation and the Small and Medium Sized Firms,* Kluwer-Nijhoff, Boston MA.

Rubenstein, J.M. (1988), 'Changing Distribution of American Motor-Vehicle-Parts Suppliers' *The Geographical Review,* vol. 78(3), pp. 288-298.

Sabel, C. (1989), 'Flexible specialization and the re-emergence of regional economies', in Hirst, P. and Zeitlin, J. (eds), *Reversing Industrial Decline? Industrial Structure and Policy in Britain and Her Competitors,* Berg, Oxford, pp. 17-70.

Sabel, C. (1993), 'Studied Trust: Building new forms of cooperation in a volatile economy', *Human Relations,* vol. 46, pp. 1133-1170.

Sadler, D. (1997), 'The Role of Supply Chain Management in the 'Europeanization' of the Automobile Production System', in Lee, R. and Wills, J. (eds), *Geographies of Economies,* Arnold, London.

Salais, R. and Storper, M. (1993), *Les mondes de production, Enquête sur l'identité économique de la France,* Editions de l'Ecole des Hautes Etudes en Sciences Sociales, Paris.

Santos, M. (1979), *The Shared Space,* Methuen, London.

Saxenian, A. (1994), *Regional Advantage. Culture and Competition in Silicon Valley and Route 128,* Harvard University Press, Cambridge and New York.

Sayer, A. (1985), 'Industry and space: a sympathetic critique of radical research', *Environment and Planning D: Society and Space,* vol. 3, pp. 3-29.

Sayer, A. and Walker, R. (1992), *The New Social Economy. Reworking the Division of Labor,* Blackwell, Cambridge, MA.

Schall, H. (1959), 'Die chemische Industrie Deutschlands. Unter besonderer Berücksichtigung der Standortfrage', *Nürnberger Wirtschafts- und Sozialgeographische Arbeiten*, (Band 2), Nürnberg.

Schamp, E.W. (1995), 'The German automobile production system going european', in Hudson, R. and Schamp, E. (eds), *Towards a New Map of Automobile Manufacturing in Europe?*, Springer Verlag, Berlin.

Schamp, E.W. (1996), 'Transition to a regional labour market on the former border between East and West Germany: an evolutionary approach to changes in the Coburg/Kronach region', in Van der Knaap, G.A. and Wever, E. (eds), *Industrial Organization: The Firm and its Labour Market*, Netherlands Geographical Studies No. 207, pp. 10-27.

Schamp, E.W. (1997), 'Raumliche Konzentration, ökonomische Kompetenz und regionale Entwicklung. Das Beispiel der oberfrankischen Autozulieferindustrie', *Erdkunde*, vol. 51, pp. 230-243.

Schoenberger, E. (1991), 'The corporate interview as a research method in economic geography', *Professional Geographer*, vol. 43, pp. 180-189.

Schulhoff, E. (1997), 'Die kraft zur erneuerung', *TM Business*, 34/97, p. 16.

Schumann, M., Baethge-Kinsky, V., Kuhlmann, C. and Neumann, U. (1994), *Trendreport Rationalisierung. Automobilindustrie, Werkzeug-maschinenbau, Chemische Industrie,* Edition Sigma Rainer Bohn, Berlin.

Schumpeter, J. (1939), *Business Cycles,* New York, McGraw-Hill.

Schumpeter, J.A. (1911/1934), *The Theory of Economic Development,* Oxford University Press, Oxford.

Scott, A. and Storper, M. (1992), 'Industrialization and regional development', in Storper, M. and Scott, A. (eds), *Pathways to Industrialization and Regional Development,* Routledge, London.

Scott, A.J. (1983), 'Industrial organization and the logic of intra-metropolitan location: 1. Theoretical considerations', *Economic Geography*, vol. 59(3), pp. 233-250.

Scott, A.J. (1986), 'Industrialization and urbanization: a geographical agenda', *Annals of the Association of American Geographers*, vol. 76, pp. 25-37.

Scott, A.J. (1988a), 'Flexible accumulation and regional development', *International Journal of Urban and Regional Research*, vol. 12, pp. 171-186.

Scott, A.J. (1988b), *New Industrial Spaces: Flexible Production Organization and Regional Development in North America and Western Europe,* Pion, London.

Scott, A.J. (1988c), *Metropolis. From the division of labor to urban form,* University of California Press, Los Angeles.

Scott, A.J. and Storper, M. (1989), 'The geographical foundations and social regulations of flexible production complexes', in Wolch, J. and Dear, M. (eds), *The Power in Geography: How Territory Shapes Social Life,* Unwin Hyman, Boston, pp. 21-40.

Scully, J.I. and Fawcett, S.E. (1994), 'International Procurement Strategies: Challenges and opportunities for the small firm', *Product and Inventory Management Journal*, Second Quarter, pp. 39-46.

Semlinger, K. (1993), 'Small Firms and Outsourcing as Flexibility Reservoirs of Large Firms', in Grabher, G. (ed), *The Embedded Firm. On the Socioeconomics of Industrial Networks*, Routledge, London.

Semlinger, K. (1995), 'Industrial Policy and Small-Firm Cooperation in Baden-Wurttemberg', in Bagnasco, A. and Sabel, C.F. (eds), *Small and Medium-size Enterprises*, Pinter, London.

Sengenberger, W. Loveman, G.W. and Piore, M.J. (1990), *The Re-emergence of Small Enterprises: Industrial restructuring in industrialised countries*, International Labour Organisation, Geneva.

Sforzi, F. (1989), 'The geography of industrial districts in Italy', in Goodman, E., Bamford, J. and Saynor, P. (eds), *Small Firms and Industrial Districts in Italy*, Routledge, London, pp. 153-173.

Shapira, P. (1992), 'Modernizing small manufacturers in Japan: the role of local public technology centers', *Journal of Technology Transfer*, Winter, pp. 40-57.

Sheard, P. (1983), 'Auto production systems in Japan: organisational and locational features', *Australian Geographical Studies*, vol. 21, pp. 49-68.

Simmie, J. (ed) (1997), *Innovation, Networks and Learning Regions?* Regional Policy and Development Series 18, Jessica Kingsley Publishers and Regional Studies Association, London and Bristol PA.

SIND (1990), Leverantörer till fordonsindustrin. (Suppliers to the Automotive Industry), *Rapport 1990:2*, Statens Industriverk, Stockholm.

Sjöstrand, S.E. (1992), 'On the Rationale behind 'Irrational' Institutions', *Journal of Economic Issues*, vol. 26(4), pp. 1007-1039.

Small and Medium Entreprise Agency (SMEA), (ed.) (1993), *Chusho kigyo shisaku soran*, Chusho Kigyo Sogo Kenkyu Kiko, Tokyo.

Small and Medium Entreprise Agency (SMEA), (ed.) (1998), *Chusho kigyo hakusho*, Okurasho Insatsu kyoku, Tokyo.

SME Finance Corporation of Japan (1986), *SME Financing in Japan*, October.

Smircich, L. and Stubbart, C. (1985), 'Strategic Management in the Enacted World', *Academy of Management Review*, vol. 10(4), pp. 724-736.

Sölvel, Ö. and Zander, I. (1991), 'Strategies for Global Competitive Advantage - the Home Based MNE vs. the Heterachical MNE', *Papers and Proceedings from the 17th Annual EIBA Conference*, Copenhagen Business School, Copenhagen.

Sölvell, Ö., Zander I. and Porter. M. (1991), *Advantage Sweden*, Norstedts, Stockholm.

Sriram, V. and Sapienza, H.J. (1991), 'An empirical investigation of the role of marketing for small exporters', *Journal of Small Business Management*, vol. 29(4), pp. 33-43.

Staber, U. 1997, 'Specialization in a Declining Industrial district', *Growth and Change,* vol. 28, pp. 475-495.

Stacey, R (1993), 'Strategy as Order Emerging from Chaos', *Long Range Planning,* vol. 26(1), pp. 10-17.

Stacey, R.D. (1996), *Complexity and Creativity in Organizations,* Berret-Koehler, San Francisco CA.

Statistics Bureau, Management and Coordination Agency (1990), *Japan Statistical Yearbook.*

Statistisches Bundesamt (1997), *Fachserie 4: Produzierendes Gewerbe. Reihe 4.1.1: Beschäftigung, Umsatz und Energieversorgung der Unternehmen und Betriebe im Bergbau und im Verarbeitenden Gewerbe 1996*t, Metzler-Poeschel, Stuttgart.

Stevenson, H.H. and Jarillo, C. (1990), 'A paradigm of entrepreneurship: entrepreneurial management.' *Strategic Management Journal,* vol. 11, Special Issue on Corporate Entrepreneurship, pp. 17-28.

Storey, D. J. and Johnson, S. (1986), 'Job generation in Britain: a review of recent studies', *International Small Business Journal,* vol. 4(4), pp. 29-47.

Storper, M. (1992), 'The limits to globalization: technology districts and international trade', *Economic Geography,* vol. 68, pp. 60-93.

Storper, M. (1995), 'The resurgence of regional economies, ten years later: the region as a nexus of untraded interdependencies', *European Urban and Regional Studies,* vil. 2(3), pp. 191-221.

Storper, M. (1997a), 'Regional economies as relational assets.' in Lee, R. and Wills, J. (eds), *Geographies of Economies,* Arnold, London, pp. 248-259.

Storper, M. (1997b), *The Regional World. Territorial Development in a Global Economy,* The Guilford Press, New York and London.

Storper, M. and Harrison, B. (1991), 'Flexibility, hierarchy and regional development: the changing structure of industrial production systems and their forms of governance in the 1990s', *Research Policy,* vol. 20, pp. 407-422.

Storper, M. and Walker, R. (1989), *The Capitalist Imperative: Territory, Technology, and Industrial Growth,* Basil Blackwell, Oxford.

Stralkowski, M. C., Klemm, C. R. et al (1988), 'Partnering Strategies: Guidelines for Successful Customer-Supplier Alliances', *National Productivity Review,* Autumn, pp. 308-317.

Streck, W.R. (1984), *Chemische Industrie: Strukturwandlungen und Entwicklungsperspektiven,* Duncker & Humblot, Berlin.

Takada, R. (1989), *Gendai chusho kigyo no kozo bunseki: koyo hendo to aratana niju kozo,* Tokyo Shinpyoron, Tokyo.

Takeuchi, A. (1990), 'Nissan Motor Company', in de Smidt, M. and Wever, E. (eds), *The Corporate Firm in a Changing World Economy,* Routledge, London.

Tarrow, S. (1996), 'Making social science work across space and time: a critical reflection on Robert Putnam's *Making Democracy Work*', *American Political Science Review,* vol. 2, pp. 389-397.

Bibliography 343

Taylor M. (1983), 'Technological change and the segmented economy', In Gillespie, A. (ed.) , *Technolofical Change and Regional Development*, Pion, London, pp. 73-88.

Taylor, M. (1984), 'Industrial geography and the business organisation', in Taylor, M. (ed.), *The Geography of Australian Corporate Power*, Croom Helm Australia, Sydney, pp. 1-13.

Taylor, M. (1987), 'Technological change and the business enterprise', in Brotchie J., Hall, P. and Newton, P. (eds), *The Spatial Impact of Technological Change*, London, Croom Helm, pp. 208-228.

Taylor, M. (1995), 'The business enterprise, power and patterns of geographical industrialisation,' in Conti, S., Malecki, E.J. and Oinas, P. (eds), *The Industrial Enterprise and Its Environment*, Avebury, Aldershort, pp. 99-122.

Taylor, M. (1996), 'Industrialisation, enterprise power and environmental change: and exploration of concepts', *Environment and Planning A*, vol. 28, pp. 1035-1051.

Taylor, M. (1999), 'The Small Firm as a Temporary Coalition.' *Entrepreneurship and Regional Development*, 11(1), pp. 1-19.

Taylor, M. and Conti, S. (eds) (1997), *Interdependent and Uneven Development: Global-Local Perspectives*, Ashgate, Aldershot.

Taylor, M. and Thrift, N.J. (1982a), 'Industrial linkage and the segmented economy: 1. Some theoretical proposals', *Environment and Planning A*, vol.14, pp. 1601-1613.

Taylor, M. and Thrift, N.J. (1982b), 'Industrial linkage and the segmented economy: 2. An empirical reinterpretation', *Environment and Planning A*, vol. 14, pp. 1615-1632.

Taylor, M. and Thrift, N.J. (1983), 'Business organisation, segmentation and location', *Regional Studies*, vol. 17(6), pp. 445-465.

Teece, D. (1992), 'Competition, cooperation, and innovation: organizational arrangements for regimes of rapid technological progress', *Journal of Economic Behavior and Organization*, vol. 18, pp. 1-25.

Tenneco Corporation [www.tenneco.com].

Terazono, E. (1993), 'Feeling the pinch: The death-knell is sounding for many of Japan's subcontractors', *Financial Times*, London, November 23.

The Economist (1994), 'MITI's identity crisis. Is small beautiful?', January 22, pp. 65-66.

The Economist (1998), 'The decline and fall of General Motors', October 10, pp. 92-97.

Thompson, J. D. (1967), *Organization in Action*, McGraw-Hill, New York.

Thorelli, H.B. (1986), 'Networks: between markets and hierarchies', *Strategic Management Journal*, vol. 7, pp. 37-51.

Tödling, F. (1994), 'The uneven landscape of innovation poles: local embeddedness and global networks.' in Amin, A. and Thrift, N.J. (eds),

Globalization, Institutions and Regional Development in Europe, Oxford University Press, Oxford.

Tolomelli, C. (1990), 'Policies to support innovation in Emilia-Romagna: experiences, prospects and theoretical aspects', in Alderman, N., Ciciotti, E. and Thwaites, A. (eds), *Technological Change in a Spatial Context: Theory, Empirical Evidence and Policy*, Springer-Verlag, Berlin/Heidelberg.

Törnqvist, G. (1963), 'Studier i industrilokalisering', *Meddelanden No.153*, Geografiska Institutionen vid Stockholms Universitet, Stockholm.

Toyoda, S. (1941), *Sangy kokusaku to chusho kigyo*, Tokyo.

Tsoukas, H. (1998), 'Introduction: Chaos, Complexity and Organization Theory', *Organizations*, vol. 5(3), pp. 291-313.

Tsuchiya, M. and Miwa, Y. (eds) (1989), *Nihon no chusho kigyo*, Tokyo daigaku shuppankai, Tokyo.

Van de Ven, A.H. (1993), 'The development of an infrastructure for entrepreneurship.' *Journal of Business Venturing*, vol. 8, pp. 211-230.

Varaldo, R. and Ferrucci, L. (1996), 'The evolutionary nature of the firm within industrial districts', *European Planning Studies*, vol. 4(1), pp. 27-34.

Vatne, E. (1995), 'Local resource mobilization and internationalization strategies in small and medium sized enterprises', *Environment and Planning A*, vol. 27(1), pp. 63-80.

VCI (1993), *Chemiewirtschaft in Zahlen. Ausgabe 1993*, 35th edition, Verband der Chemischen Industrie, Frankfurt/Main.

VCI (1994a), *Firmenhandbuch Chemische Industrie 1994/95*, 16th edition, ECON, Düsseldorf.

VCI (1994b), *Chemiewirtschaft in Zahlen. Ausgabe 1994*, 36th edition, Verband der Chemischen Industrie, Frankfurt/Main.

VCI (1997), *Chemiewirtschaft in Zahlen. Ausgabe 1997*, 39th edition, Verband der Chemischen Industrie, Frankfurt/Main.

VLI (1988), *Jahresbericht 1987*, Verband der Lackindustrie, Frankfurt/Main.

VLI (1994), *Jahresbericht 1993*, Verband der Lackindustrie, Frankfurt/Main.

VLI (1997), *Jahresbericht 1996*, Verband der Lackindustrie, Frankfurt/Main.

Vogel F. (1975), *Probleme und Verfahren der numerischen Klassifikation. Unter besonderer Berücksichtigung von Alternativmerkmalen*, Vandenhoeck & Ruprecht, Göttingen.

Volvo Group [www.volvo.com].

Walbro Corporation [www.walbro.com].

Watson, T.J. (1994), 'Entrepreneurship and professional management: a fatal distinction.' *International Small Business Journal*, vol. 13(2), pp. 34-46.

Weick, K.E. (1976), 'Educational organizations as loosely coupled systems', *Administrative Science Quarterly*, vol. 2 (March), pp. 1-19.

Weick, K.E. 1979 (1969), *The Social Psychology of Organizing*, Addison-Wesley, Reading MA.

Wheatley, M. (1992), *Leadership and the New Science*, Barret-Koehler, San Francisco.

Whittaker, D.H. (1997), *Small Firms in the Japanese Economy*, Cambridge University Press, Cambridge.

Wicken, O. (1994), 'Entreprenørskap i Møre og Romsdal: Et historisk perspektiv', *STEP Report* 17/94, STEP Group, Oslo.

Williamson, O. (1975), *Markets and Hierarchies. Analysis and Antitrust Implications. A Study in the Economics of Internal Organization*, Free Press, New York.

Williamson, O.E. (1979), 'Transaction cost economics: The governance of contractual relations', *Journal of Law and Economics*, vol. 22, pp. 223-241.

Williamson, O.E. (1985), *The Economic Institutions of Capitalism*, Free Press, New York.

Womack, J.P., Jones, D.T. et al (1990), *The Machine That Changed the World*, Rawson Associates, New York.

Wrong, D. (1995), *Power: Its Forms, Bases, and Uses*, Transaction Publishers, New Brunswich and London.

Yeung, H. (1998a), 'Capital, state and space: contesting the borderless world', *Transactions of the Institute of British Geographers, New Series*, vol. 23, pp. 291-309.

Yeung, H. (1998b), 'The social-spatial constitution of business organizations: a geographical perspective', *Organizations*, vol.5 (1), 101-128.

Ylinenpää, H. (1999), 'Competence Management and Small Firm Performance', in Johannisson, B. and Landström, H. (eds), *Images of Entrepreneurship and Small Business - Emergent Swedish Contributions to Academic Research*, Studentlitteratur, Lund, pp. 216-238.

Index

Printed and bound by CPI Group (UK) Ltd, Croydon, CR0 4YY

22/10/2024

01777620-0002